edexcel
advancing learning, changing lives

Student Book and CD-ROM

Course Companion
360Science

GCSE Science
Foundation and Higher Tier

Pauline Anderson   Alan Philpotts

David Horrocks   Ian Roberts

Sue Jenkin   Julia Salter

Vanessa Love   Richard Shewry

A PEARSON COMPANY

Edexcel
190 High Holborn
London WC1V 7BH
UK

ISBN-10 1-84690-158-8 **ISBN-13** 978-1-84690-158-4

Printed and bound in China GCC/03
Prepared for Edexcel by Starfish Design, Editorial and Project Management Ltd
Project management by Heather Morris
Illustrated by Peters and Zabransky (UK) Ltd
We are grateful to Oxford Designers and Illustrators Ltd for permission to reproduce the illustrations on pp 18, 27, 28, 29, 30, 38, 40, 47, 50, 53, 56, 62, 66, 71

Summary of Licence Terms (full licence on disk)

Yes, You can:
1. use the Software on your own personal computer as a single individual user.

No, You cannot:
1. copy the Software (other than making one copy for back-up purposes);
2. alter the Software, or in any way reverse engineer, decompile or create a derivative product from the contents of the database or any software included in it (except as permitted by law);
3. include any of the Software in any other product or software materials;
4. rent, hire, lend or sell the Software to any third party;
5. copy any part of the documentation except where specifically indicated otherwise;
6. use the Software in any way not specified above without the prior written consent of Edexcel.

Type of Licence:
Subject to your acceptance of the full licence terms, Edexcel grants a non-exclusive, non-transferable licence to use the Software for personal, non-commercial use.

Warranty:
Edexcel grants a limited 90 day warranty that CD-ROM on which the Software is supplied is free from defects in material and workmanship.

Governing Law:
The licence is governed and construed in accordance with English law.

Contents

This new course will help you to understand the science around you that affects everyday life and to develop your own practical scientific skills and knowledge. It includes Biology, Chemistry and Physics and each science has four interesting and useful topics that also form part of the separate GCSE Biology, GCSE Chemistry and GCSE Physics qualifications.

Biology Topic 1 Environment

This topic focuses on how people interact with the environment and covers many important issues that we need to think about for the future of our planet.

- When a species becomes extinct, is this caused by predators, or are human beings to blame?
- What causes desertification and deforestation?
- How can you measure energy losses in food chains to find out if producing grain or meat is more efficient?
- Organic food is becoming more popular but is it really better for you or for the environment?
- How is selective breeding of animals and plants (which people have done for thousands of years) different from making genetically modified organisms?
- How does Charles Darwin's theory of evolution fit with new evidence from DNA?

Biology Topic 2 Genes

This topic is all about genes and DNA and looks at some of the exciting developments in research.

- Is natural variation only caused by mixing the genes from two parents?
- What is the effect of nurture as well as nature on a person's growth?
- Did you know that many plants have only one parent and are clones?
- What do you know about modern clones such as Dolly, or Tracey the transgenic sheep?
- What are the causes of genetic disorders and what new therapies are being developed?
- What might the Human Genome Project mean for further research?
- What are the ethical issues raised by gene therapy and designer babies?

Biology Topic 3 Electrical and chemical signals

Our bodies have a fast electrical and a slower chemical communication system. This topic explains how these two systems work.

- Do you know how messages from our sense receptors are carried along our nerves to our spinal column and our brain?
- What are the different types of nerves and the different senses?
- How do reflexes like ducking and blinking help to keep us healthy and safe?

- You will have heard of the hormones insulin, testosterone and oestrogen, but do you know just how many others there are and what they do?
- Female hormones can be used to control fertility, for example, in contraceptive pills, but how do these work?
- What are the ethical issues involved in IVF?

Biology Topic 4 Use, misuse and abuse

In this topic you will learn how our bodies constantly defend us from attack by bacteria and viruses and how modern drugs can help.

- How are disease-causing microorganisms transmitted?
- How do our immune systems deal with them?
- Have you heard that some bacteria are becoming immune to antibiotics?
- What are the different types of drugs and how do they affect our bodies?
- Are all legal drugs safe?
- What do you know about the side effects of different drugs and why it can be so difficult to give them up?

Chemistry Topic 5 Patterns in properties

This topic introduces the core tools of chemistry, including the periodic table, chemical formulae and equations and shows how they can be used practically.

- Could you be a forensic scientist and use flame tests to identify some metals in their compounds to catch criminals at a crime scene?
- Find out about the alkali metals (watch out for some violent reactions here), the halogens, the noble gases, and the transition metals.
- Have you noticed that when some chemical reactions take place, the container becomes hot?
- Can you explain why this is so and what is happening?
- Do you think that the opposite can also happen?

Chemistry Topic 6 Making changes

There are many opportunities for practical work in this topic, for example, growing crystals and looking at them under a microscope.

- Do you know how some metals are extracted using carbon and that our ancestors first did this by accident in the stone age?
- Have you ever wondered why cakes rise when they are being cooked in the oven?
- How can you test for different gases and what are the different ways to collect samples of them?
- Have you ever heard of compounds such as ethanoic acid, sodium chloride, phosphoric acid and citric acid?
- What chemical substances are found in common foods and drinks?

Chemistry Topic 7 There's one Earth

This topic focuses on the resources we use, how we obtain them and what impact our consumption is having on the Earth.

- What exactly is global warming, how is it caused, and just how much of a problem is it?
- Which is the best fuel to use for different purposes?
- What will be the fuel of the future when petrol runs out?
- How do we obtain petrol anyway?
- Have you heard stories about people being killed due to faulty gas appliances?
- Did you know that sea water is the source of lots of different useful chemical substances?

Chemistry Topic 8 Designer products

This topic is about being close to the cutting edge of new developments in chemistry.

- How were new materials such as Lycra™, Thinsulate™ and Kevlar™ developed and what are their uses?
- How do spectacles that you can sit on without damaging them 'remember' their shape?
- You have probably heard the expression 'nanotechnology' but what is it all about?
- Find out about how beers, wines and spirits are made all using a very simple, basic chemical reactions.
- What is 'intelligent packaging' and how does it help keep food really fresh?
- What do mayonnaise, paints, hair gels and shampoos have in common?

Physics Topic 9 Producing and measuring electricity

Many of the things we do would be impossible without electricity. This topic gives you a chance to discuss the tremendous changes that electricity has made for the modern world and think about what might happen next.

- How many different sources of electricity do you know about?
- What makes them suitable for the many different jobs we want them to do?
- How can we control, for example, how long the shutter on a digital camera stays open?
- What might happen next in telephone technology, or in the development of fast trains, or the improved processing speed of computers?

Physics Topic 10 You're in charge

Electricity is not dangerous if treated with respect. Basic safety precautions allow us to use it in so many applications that you don't normally think about.

- Many people think that the electric motor was the greatest invention of all time – would you agree?
- Learn how it works and try to make one – it can be quite simple and great fun!
- What role does electricity play in medicine?
- How can it measure and control the beating of hearts, monitor our brains and move the muscles of paralysed people?
- How possible is it to limit the amount of electrical energy we use to save money as well as doing our bit towards saving the planet?

Physics Topic 11 Now you see it, now you don't

There are many different types of waves. We can detect visible light using our eyes, sound with our ears and infra-red with our skin but most of the waves we use are invisible. We only know they are there by things they do.

- How do X-rays affect photographic paper?
- How do radio waves help us to communicate?
- We all know microwaves can cook food, but what have they got to do with weather forecasting?
- How can we see inside luggage at airports?
- What do X-rays and seismic waves have in common?
- How much do we know about the harmful effects to the human body of over-exposure to some types of wave?

Physics Topic 12 Space and its mysteries

This topic explores what we know about our nearest neighbours, including all the planets as well as comets and asteroids. By the end, you should be able to separate those things which are true science from those which, at present, are science fiction.

- How we can escape Earth's gravity and survive the unusual conditions of weightlessness, temperature variations, radiation and lack of a suitable atmosphere in space?
- We have been able to send unmanned probes to distant planets but what obstacles will need to be overcome before humans can follow them?
- Are alien life forms a possibility?
- If so, on what type of planet might they live and how might we make contact with them?
- Where did we come from? What do you know about the Big Bang theory?
- Has our Sun always been the same and how long will it last?
- Are all stars like our Sun?

How Science Works

How science works is a part of all the science you study. It is both about understanding the science you meet in everyday life and about how you carry out your own scientific investigations. In studying the twelve topics in GCSE Science your teacher will help you to develop knowledge and understanding of how science works in that context.

Understanding how science works will help you to:

- develop practical skills
- identify the questions that science can and cannot answer
- understand how scientists look for the answers
- evaluate scientific claims by judging the reliability and validity of the evidence
- question scientific reports in the media
- communicate your own findings clearly
- consider scientific findings in a wider context – recognising the difference between a fact and a theory
- make informed judgements about science and technology, including any ethical issues that may arise.

There are four aspects to developing your understanding of how science works:

1   Practical and enquiry skills

Scientists have to plan investigations and experiments to find evidence to support their ideas or put forward an answer to a scientific problem. They might want to:

- test a scientific idea
- answer a scientific question
- solve a scientific problem.

The important part of designing practical experiments is to ensure that the tests are fair, that reliable data is collected and the investigation is safe.

2   Data, evidence, theories and explanations

Scientists collect and analyse data. This might be primary data from their own experiments or secondary data from someone else's, often reported on the internet. If the data is secondary, they must be very careful to say where the data came from, by giving the name of the website, etc. Scientists will always look at their data and how they collected it to make sure it is safe and accurate.

- Was the method used the right one?
- Are the data accurate enough?
- Are the data valid and reliable so that they can be used as evidence?

If scientists are confident about their data, these could then be used to develop an argument to say what the evidence shows, and would be used to draw a conclusion. It is very important that any conclusion is well supported by the evidence.

Scientists often use models to try to predict what will happen in the future. A model allows a scientist to try out ideas and see what might happen if something is changed in the model. Global warming (see pp. 52) is an example of where models are very important in trying to predict future events.

Scientists use data from experiments and models to create theories that explain events or phenomena. Scientific theories are not just guesswork but are based on reliable evidence.

3   Communication skills

Scientists communicate with each other to share ideas and to see how others feel about their ideas. They communicate with the general public so that everyone knows what they have found out and how those ideas might be used.

Scientific ideas and information need to be analysed, interpreted and questioned. This is how scientists look at each other's work. If another scientist cannot carry out the experiment someone has tried and get the same results, the scientist would be very suspicious. In 2006, a South Korean scientist's work on stem cells was shown to be based on false data.

Data and information need to be converted into a form that other people can understand. Scientists often use ICT tools such as spreadsheets to help them do this. Spreadsheet data can be converted into graphs which are often easier to understand. Graphs could then be used in a slide presentation to communicate to other people.

4   Applications and implications of science

Many scientific and technological advances have benefits, drawbacks and risks. Often decisions about science and technology are difficult, because they raise ethical issues, and these decisions may have social, economic and environmental consequences. For example, should any woman who wants it be given IVF treatment? Adriana Iliescu gave birth after fertility treatment. She was 66 when her daughter was born so when Adriana is 80 her daughter will be 14. What issues does this raise? See page 32.

At the end of the course – what qualification will I achieve?

You are working towards gaining a GCSE Science qualification and during the course there will be opportunities to find out how well you understand how science works. Your teacher will ask you questions in class or homework that will help you to assess whether you understand and know the science in the twelve topics. In addition to the day-to-day help from your teacher in assessing your progress, you will be required to take assessments set by Edexcel. It will be Edexcel that will award your final grade for GCSE Science.

What will I need to gain a good GCSE Science grade?

Are Biology, Chemistry and Physics topics equally important?
Yes, each accounts for a third of the final GCSE Science grade.
You will be assessed in three ways as shown here:

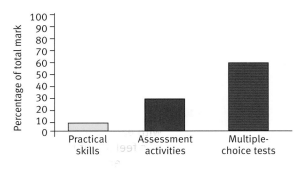

Assessment of practical skills (10% of total marks)

Your teacher will assess three skills:
- how well you follow instructions (probably given on a worksheet)
- how good you are at collecting data safely, accurately and reliably
- your ability to present results in tables with headings and units.

Your teacher may gave you further guidance on when and where they will be watching you carry out some work in the laboratory. There are more details on the CD to explain what your teacher will be looking for in the observation. The main aim for you is to try to improve your practical skills during the course so that you achieve the highest mark possible in the three skill areas.

Assessment activities (30% of total marks)

These activities are set by Edexcel. Your teacher will decide when you will take one of these assessments during the course. There is an assessment activity for each of the twelve topics in GCSE Science. It will depend on the plan for your course which ones you are asked to take on completing a particular topic. In some instances you may take more than one in Biology, Chemistry or Physics. The highest score from each subject will account for 10% of your total score – making a possible 30% in total.

The assessment activities will test your understanding of how science works in a particular topic. Your teacher will require you to work on your own and you will have 45 minutes to complete the task.
Your teacher will mark the assessment activities. When you get your marked task back, try to identify the type of questions you have done well on and think about the ones you would need to prepare more thoroughly to improve future scores.

It may be that you would be asked to plan an experiment, so what would you need to consider?

Planning gives you the opportunity to make your own worksheet for the experiment. You will need to think about things like:
- what you will measure and how often
- what instruments you will use
- in what order you will do the various activities
- what you will control so as not to affect the result unfairly
- what safety precautions you will take.

Analysing may include plotting graphs to find patterns in the data provided in the task and drawing conclusions to generalise your findings.

Evaluating involves judging whether the evidence given to you in the task is based on reliable results or suggesting what improvements could have been made. For example, you may have to consider the implications of evidence given to you on reducing the number of cars on the roads to prevent global warming.

Multiple-choice tests (60% of total marks)

There will be six tests to cover the twelve topics – two for biology, two for chemistry and two for physics. Each test will cover two topics and you are allowed 20 minutes.

After discussion with your teacher you can choose whether to take the Foundation Tier test (questions 1–24) or the Higher Tier test (questions 17–40). You can take different levels in different tests if you think that will give you the best score.

The text in this book in green is for Higher Tier students only.

Edexcel will mark these tests and you will be sent the results. You will find it useful to keep your test results in a safe place to know how your score is building up. The tests are available in November, March and June. Your teacher will let you know on which date you will take the tests. Look on the CD to see a sample module test paper. Each test counts for 10% of the total marks. If you take a test more than once, the highest score will count.

And finally

Your final GCSE Science grade will be awarded by Edexcel based on your total score from:
Practical Skills – out of 10%
Assessment Activities – out of 30%
Multiple-choice Tests – out of 60%

This makes a total out of 100%. Throughout the course your teacher will have discussed with you a target grade. Your challenge is to take the opportunity to meet your target and achieve the highest grade possible.

Preparation is the key to success. Your grades are your responsibility!

Make the most of your lessons by preparing for them and following the advice of your teacher. There will be many opportunities during the course to find out about your progress, either with help of your teacher or by testing yourself. Try the questions on the CD and fill in the self-assessment forms to check your progress.

Science lessons

Before a science lesson
- Glance through your notes.
- Write down three SMART targets for success.
- Check that you have your pen, pencil, calculator and books.

What are SMART targets?
Small
Measurable
Achievable
Realistic
Time limited
'I will learn 15 vocabulary words for genetics on Monday' is better than 'I will do some Biology revision next week'.

During the lesson
- Stay on task.
- Make your notes clear.
- Ask sensible questions.

After the lesson
- Make a short summary of the lesson.
- Write down questions to ask in your next lesson.
- Practise questions from the book.

Preparing for assessment activities and multiple-choice tests

Make summaries of your notes in your own words.

Review and grade your notes.
1. I can do and I know this.
2. I understand this but I don't know this yet.
3. I don't understand this and I don't know this yet.

Ask for help with the work you don't understand.

Learn: your vocabulary, equations, how things work, labels on diagrams, units.

Timetable your revision: start early – do 20 minutes every other day – spend more time on the topics you don't understand and don't know but ensure that you look at everything.

Make memory aids for yourself if you find these helpful. For example, the colours of the rainbow Red, Orange, Yellow, Green, Blue, Indigo and Violet can be remembered by Richard Of York Gave Battle In Vain.

When you have finished a topic practise exam questions on that topic – look on the CD for some examples.

Practise questions
- Write specimen answers and check them against the mark scheme.
- If you get it wrong try again after some more revision.
- Pretend you are the examiner and write a mark scheme for the question.

The day before a test
- Check your pens etc work and are in your bag.
- Get enough sleep.
- Don't cram, just read through your summaries.

Taking the test
- Check how long the test is and work out where you need to be by the half-way stage.
- Read the instructions and the introduction to each question carefully.
- Look at how many marks are given for each part of each question.
- Make a point for each mark.
- If you get stuck, move on to the next question but remember to come back to it if you have time.
- Write in the space provided or use extra sheets.
- Make your writing legible and in dark blue or black pen – the examiner has to be able to read it if he is to mark it.
- Use the correct vocabulary.
- Read through your work carefully and critically at the end.

Doing calculations
- Write down the equation.
- Make it clear what values you are substituting.
- Write down your working out.
- Set out your working neatly.
- Remember to add the units.

Environment

1 Competition for resources

Have you ever wondered?

Why don't food chains go on forever?

Key facts

- Animals compete with each other for resources like food.
- Animals can compete between species and within a species.
- The strongest, best adapted animals get to eat most of the food, so they survive and breed at the expense of less well adapted animals.
- Predators kill and eat other animals.
- Scavengers are animals that eat what is left after predators have eaten what they need of their prey; they also eat other dead animals.
- Plants and animals are **interdependent**. One cannot live without the other.
- Changes in the numbers of one organism affect the numbers of other organisms.
- **Population** changes in predators lag behind changes in the population of their prey.

Definitions

- **Interdependent** a change in one organism causes a change in another
- **Population** the total number of a single species living in an area

Look on the CD for more exam practice questions

❶ Examiner's tips

- You need to be able to explain the shape of predator/prey graphs.
- You need to be able to explain why population changes in predators lag behind changes in population of their prey.
- You should understand why computer models have advantages and disadvantages.

Can you answer these questions?

1. What do scavengers do?
2. Explain what will happen to the numbers of predators if the number of their prey falls.
3. Explain why populations of better adapted animals increase.
4. Why do animals compete with each other?

Did you know?

- Charles Darwin described life as a struggle for survival.

② Our influence on the environment

Have you ever wondered?

How do different organisms make different changes to solve the same environmental problem?

Key facts

- The population of humans has increased.
- This increase in population puts pressure on resources.
- Every person needs water, food and somewhere to live.
- Forest is burned and cleared to make more room for farming.
- Burning fuel for electricity generation and transport produces carbon dioxide.
- Carbon dioxide is a greenhouse gas, and contributes to global warming.
- Over-using land can make it less fertile, which can result in desertification.
- Cutting down rainforest is an example of deforestation.

Definitions

- **Environment** the place where an organism lives, e.g. air, water, soil, other living things

❶ Examiner's tips

- You should know about the probable consequences of global warming and some things we can all do to help the **environment**.
- You need to be able to explain the shape of graphs showing human population growth and the link between human population and increased carbon dioxide in the atmosphere.

Look on the CD for more exam practice questions

Can you answer these questions?

1. What has happened to the population of humans over the last 100 years?
2. What effects has this had on the environment?
3. Why is carbon dioxide a greenhouse gas?
4. Why does deforestation take place?

Did you know?

- If current predictions are correct, there will be 10 billion people on Earth by 2150

③ Chains, webs and pyramids

Have you ever wondered?
How can the Sun's energy support all life on Earth?

Key facts

- Green plants use photosynthesis to make their own food.
- Green plants are producers.
- The arrows in a food chain show the flow of energy through the food chain.
- Several food chains link together to make a food web.
- A food pyramid shows the numbers of organisms in a food chain.
- Pyramids of number show how many of each organism are found at each level of a food chain.
- Pyramids of **biomass** show the dry mass of living matter in a food chain.

Definitions
- **Biomass** mass of an organism (or population) after water is removed
- **Ecosystem** interacting group of plants and animals

Look on the CD for more exam practice questions

❶ Examiner's tips
- An **ecosystem** is made up of lots of different food chains, and these can be linked to make a food web.
- Most food chains begin with energy from the Sun, but this is not shown in a food chain diagram.
- There is a difference between pyramids of number and pyramids of biomass. A pyramid of biomass is more useful.

Can you answer these questions?
1. Name two producers.
2. Explain how the Sun's energy supports all life on Earth.
3. Explain the difference between pyramids of number and pyramids of biomass.
4. What is a food web?

Did you know?
- There are volcanic vents on the sea bed where living organisms get their energy from chemicals, not the Sun.

4 Wheat versus meat

Have you ever wondered?

Which grows more quickly – grass or cow?

Key facts

- Food chains are made up of only three or four links.
- Only about 1% of the energy falling on a wheat field is used for photosynthesis. The rest is lost though reflection and evaporation of water from plants.
- Only about 10% of the energy captured by a plant is transferred to primary consumers.
- Only about 10% of the energy eaten by a primary consumer is transferred to secondary consumers.
- Feeding cereals to animals that we then eat is a very inefficient way of making protein.
- **Organic farming** is a more natural way of producing food by avoiding the use of chemicals when growing crops.

Definitions

- **Organic farming** uses natural ways to control pests and keep soil fertile

Look on the CD for more exam practice questions

❶ Examiner's tips

- Energy is lost at every stage of a food chain.
- Food chains are short because so much energy is lost at each stage.
- Factory farming and organic farming are different ways of producing food; both have advantages and disadvantages.

Can you answer these questions?

1. How much energy falling on a wheat field is used in photosynthesis?
2. Why is it inefficient to feed cereals to animals that we then eat?
3. Explain what happens to most of the sunlight that falls onto a field of wheat.
4. Why are food chains very short?

Did you know?

- 1 m^2 of a wheat field receives 1000 MJ of energy a year from the Sun.

(5) Natural selection and evolution

Have you ever wondered?

How does natural selection 'know' how to create a new species?

Key facts

- The theory of evolution was developed by Charles Darwin.
- Fossils are important evidence for evolution.
- An important part of evolution is natural selection.
- All organisms produce many young, so the population should increase, but it doesn't, so some offspring must die.
- Organisms which have variations which give them better **adaptations** to their environment are more likely to survive and breed.
- Some characteristics are inherited, so useful characteristics may be passed on to future generations. Eventually this may give rise to a new species.
- Natural selection takes place over very long periods of time.
- If there is a big change to the environment, some groups of organisms cannot adapt and become extinct.

Definitions

- **Reproduction** propagation of the next generation
- **Adaptation** how organisms change to suit their environment better

❶ Examiner's tips

- Natural selection happens over long periods of time, from generation to generation, not within a generation.
- Natural selection depends upon variations between individual organisms within a population. Those with the best-suited variation are most likely to survive and breed.
- Charles Darwin's theory is very difficult to prove, although there is a lot of evidence to support it.

Look on the CD for more exam practice questions

Can you answer these questions?

1. What are fossils?
2. Why is variation an important part of natural selection?
3. Why do some organisms become extinct?
4. Explain how natural selection takes place.

Did you know?

- Some people think that Darwin's theory of evolution should not be taught in schools

6 Variation

Have you ever wondered?

Why did a cartoon of Charles Darwin drawn as an ape appear in a national newspaper?

Key facts

- There are millions of different species of organisms in the world.
- Scientists record these and put them into groups. This makes it easier to identify new species when they are discovered.
- The groups contain related organisms to try and show how they evolved.
- Humans and apes are in a group together called primates.
- The animal kingdom is divided into animals with backbones and those without backbones.
- Some organisms are difficult to classify: *Euglena* is an animal but it can photosynthesise like plants.
- The smallest **classification** of organisms is called a species, and even then there is variation amongst the individuals.

Definitions

- **Classification** a way to group things systematically

Look on the CD for more exam practice questions

❶ Examiner's tips

- Classification is all about trying to put organisms into groups.
- Vertebrates and invertebrates are two important groups.
- Classification is really useful when a new organism is found.

Can you answer these questions?

1. What is the name of the group to which humans belong?
2. How is the animal kingdom divided into two big groups?
3. Why is it very difficult to classify some organisms?
4. Explain why scientists put organisms into groups.

Did you know?

- About 16 000 species are currently threatened with extinction. Nearly 1000 species became extinct in 2004.

Genes

① Chromosomes, genes and DNA

Have you ever wondered?

How some organisms, including humans, can have one eye that is blue and one eye that is brown?

Key facts

- **Chromosomes** are found in the nucleus of the cell.
- During **fertilisation** chromosomes from our mother pair up with chromosomes from our father.
- This means that we have two copies of every chromosome and two copies of every gene (or a gene pair), one on each chromosome.

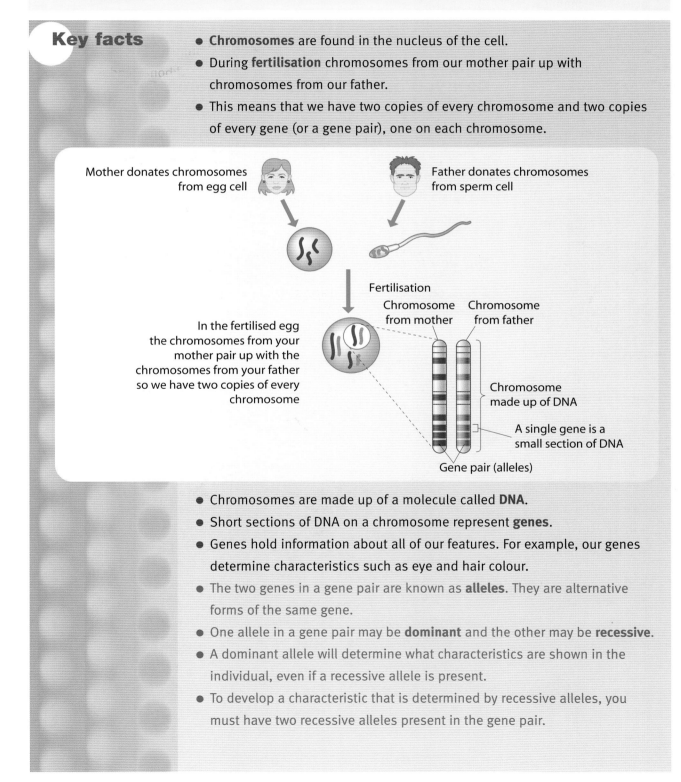

Mother donates chromosomes from egg cell

Father donates chromosomes from sperm cell

In the fertilised egg the chromosomes from your mother pair up with the chromosomes from your father so we have two copies of every chromosome

Fertilisation

Chromosome from mother   Chromosome from father

Chromosome made up of DNA

A single gene is a small section of DNA

Gene pair (alleles)

- Chromosomes are made up of a molecule called **DNA**.
- Short sections of DNA on a chromosome represent **genes**.
- Genes hold information about all of our features. For example, our genes determine characteristics such as eye and hair colour.
- The two genes in a gene pair are known as **alleles**. They are alternative forms of the same gene.
- One allele in a gene pair may be **dominant** and the other may be **recessive**.
- A dominant allele will determine what characteristics are shown in the individual, even if a recessive allele is present.
- To develop a characteristic that is determined by recessive alleles, you must have two recessive alleles present in the gene pair.

Definitions

- **Chromosome** a thread-like string of genes in the nucleus
- **DNA** deoxyribonucleic acid – the chemical code that governs cell development
- **Gene** a piece of DNA with the instructions for a particular characteristic, e.g. eye colour
- **Allele** alternative forms of the same gene, e.g. the flower colour gene in peas has alleles for white and yellow
- **Dominant** an allele that overrides all other alleles of a gene, hiding their effects
- **Recessive** an allele that is overridden by a dominant one – its effects only show when it is inherited from both parents

Look on the CD for more exam practice questions

❶ Examiner's tips

- Students often make the mistake of stating that alleles are two copies of the same gene. Alleles are *different versions* of the same gene. Think about loaves of bread on a supermarket shelf. Some are white, some wholemeal, some multigrain. They are all loaves of bread, just different versions of each other. This is the same as genes. We might inherit a recessive gene for blue eyes from our mother and a dominant gene for brown eyes from our father. They both determine the same characteristic but vary slightly – in this case in the colour of the characteristic that they give instructions for.

Can you answer these questions?

1. What is the name of the molecule that makes up genes?
2. If you inherited a recessive gene for blue eyes from your mother and a dominant gene for brown eyes from your father, what colour would your eyes be?
3. Explain the difference between a chromosome and a gene.
4. What are alleles?

Did you know?

- 50% of human DNA is identical to that of a banana.

② Variation

Have you ever wondered?

What it would be like if offspring were always identical to one of their parents?

Key facts

- Organisms that are produced by **asexual reproduction** have only one parent.
- Asexual reproduction produces **clones** which are genetically identical to the parent.

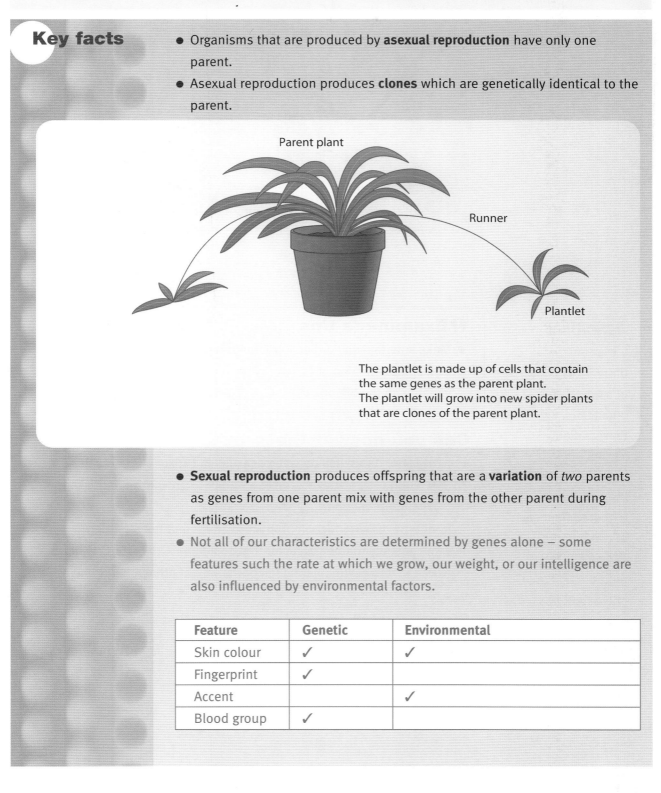

Parent plant

Runner

Plantlet

The plantlet is made up of cells that contain the same genes as the parent plant.
The plantlet will grow into new spider plants that are clones of the parent plant.

- **Sexual reproduction** produces offspring that are a **variation** of *two* parents as genes from one parent mix with genes from the other parent during fertilisation.
- Not all of our characteristics are determined by genes alone – some features such the rate at which we grow, our weight, or our intelligence are also influenced by environmental factors.

Feature	Genetic	Environmental
Skin colour	✓	✓
Fingerprint	✓	
Accent		✓
Blood group	✓	

● Environmental factors, such as lack of minerals, can also affect the health of plants including their growth.

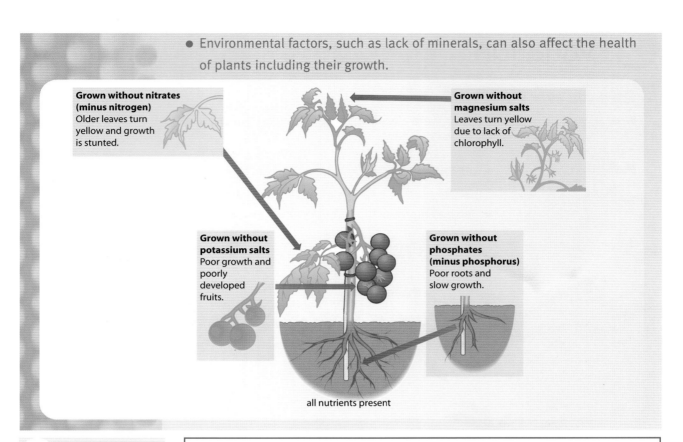

Grown without nitrates (minus nitrogen) Older leaves turn yellow and growth is stunted.

Grown without magnesium salts Leaves turn yellow due to lack of chlorophyll.

Grown without potassium salts Poor growth and poorly developed fruits.

Grown without phosphates (minus phosphorus) Poor roots and slow growth.

all nutrients present

Definitions

● **Variation** differences shown in a group of organisms, e.g. flower colour
● **Sexual reproduction** half the genes are inherited from each of two parents
● **Asexual reproduction** only one parent passes on genes
● **Clone** genetically identical plants or animals produced asexually from one parent

Look on the CD for more exam practice questions

❶ Examiner's tips

● Make sure that you understand the differences between asexual and sexual reproduction. It is only sexual reproduction that produces variation in the offspring.

Can you answer these questions?

1. What is the difference between asexual and sexual reproduction?
2. Why might it be an advantage to a farmer to grow his plants using asexual reproduction?
3. What type of reproduction produces variation in the offspring?
4. Name one environmental factor that could affect the rate at which we grow and one environmental factor that might affect plant growth.

Did you know?

● The heaviest man on record weighed 636 Kg, the same as five baby elephants.

③ Inheriting disease

Have you ever wondered?
How genetics can be used to cure diseases?

Key facts

- Some disorders, such as **cystic fibrosis** (CF), can be **inherited** from our parents.
- CF is a recessive disorder, which means that the offspring must have two recessive alleles, one from each parent.
- CF causes certain cells in the lungs and digestive system to produce large amounts of a thick, sticky mucus.
- Huntington's disease affects the brain and nervous system.
- Huntington's disease is caused by a dominant allele so only one parent needs to carry the faulty gene.

Parents' genes **Hh** **hh**

All eggs contain the normal allele.

Half of the sperm cells produced will carry the Huntington's disease allele (**H**). The other half carry the normal allele (h).

	h	h
h	hh	hh
H	**Hh**	**Hh**

50% of the children carry the dominant Huntington's disease allele (**H**) so they will have the disorder. The remaining 50% are normal.

Definitions
- **Cystic fibrosis** genetic disorder causing abnormal, sticky mucus in the lungs
- **Inheritance** the passing of genes from parents to offspring

❶ Examiner's tips
- Don't forget that we have two genes for every feature including any inherited disorders that we have. As we can only inherit one gene from each parent for a particular feature, then to inherit a recessive characteristic like CF, both parents must carry at least one faulty allele.

Can you answer these questions?
1. How many copies of the faulty allele must a person inherit to have cystic fibrosis?
2. Suggest why a person with cystic fibrosis may be underweight.
3. Explain why the parents of a cystic fibrosis sufferer may show no symptoms of the disease themselves.

Look on the CD for more exam practice questions

Did you know?
- There are about 2000 genetic disorders that are caused by a dominant allele and around 1000 caused by a recessive allele.

4 Gene therapy

Have you ever wondered?

What would happen to the world population if all diseases could be prevented or treated with genetics?

Key facts

- Gene therapy is a process that involves repairing faulty genes or replacing them with healthy ones.
- Genetic disorders such as cystic fibrosis (CF) and some types of breast **cancer** could soon be prevented or cured using gene therapy.
- Viruses or tiny lipid droplets (liposomes) can carry the healthy genes into certain cells that make up the lungs of CF sufferers.
- A DNA fingerprint is your unique barcode that distinguishes you from everyone else.
- A DNA fingerprint can be used to determine whether an individual has an inherited disease or is at a greater risk of developing a disease.
- Individuals at a greater risk of developing diseases such as breast cancer can be monitored more closely.
- **Forensic** scientists can use DNA fingerprints in many ways including identifying criminals or crime victims.

Definitions

- **Cancer** rapid uncontrolled growth of cells to form tumours which may spread
- **Forensic** using scientific knowledge to help detect crime

Look on the CD for more exam practice questions

❶ Examiner's tips

- You are born with genetic diseases but it is slightly different in the case of breast cancer. Some people are more at risk of this disease than others. This disease tends to develop later on in life as the faulty gene is more likely to be 'switched on', possibly by environmental factors such as a poor diet.

Can you answer these questions?

1. Explain how people with cystic fibrosis can be treated using gene therapy.
2. What is a DNA fingerprint?
3. Give one medical use and one use in forensic science of DNA fingerprinting.

Did you know?

- The majority of your DNA is identical to everyone else's. Only 0.1% differs between individuals and it is this varying region that is used to make a DNA fingerprint.

5 The human genome project (HGP)

Have you ever wondered?

What it would be like if you walked into a restaurant to be served by someone who looked and spoke just like you?

Key facts

- DNA is made up of smaller units called bases.
- There are 4 different types of bases in DNA. They are given the letters A, G, C and T.
- Many millions of these bases line up to form a chromosome, and short stretches of these bases make up the codes for our genes.
- The **human genome project (HGP)** has worked out the order of these bases on all of our chromosomes.
- Using the information gained from the HGP, scientists have been able to work out the codes for certain genes including some that cause disease.
- Eventually, the codes for all our genes will be known and some people are very concerned about the effects that this might have on their lives.

Definitions

- **Human genome project (HGP)** a project that worked out the sequence of genes in human DNA

❶ Examiner's tips

- There are many benefits in knowing the codes for all of our genes but there are also some concerns. Make sure that you can name at least two advantages and two disadvantages of the HGP.

Can you answer these questions?

1. What is the HGP?
2. Suggest how knowing the codes for our genes might help gene therapy.
3. Give two advantages and two disadvantages of the HGP.

Look on the CD for more exam practice questions

Did you know?

- It took 13 years to work out the order of the 3 billion bases found on our chromosomes. If we wrote down this sequence it would take up 15, four drawer filing cabinets, each stuffed with A4 paper written on both sides with the letters AGCT.

⑥ Playing with genes

Have you ever wondered?

Could we really clone dinosaurs like they did in Jurassic Park?

Key facts

Transgenic animals

- Genes can be transferred from one organism to another to produce a **transgenic organism**.
- Transgenic organisms have new or altered characteristics.
- Some transgenic organisms have human genes inserted into their cells.
- These organisms can produce milk containing human antibodies or other useful substances to treat diseases, for example cystic fibrosis.

Cloning mammals

- Scientists are able to create clones of mammals and other organisms artificially.
- Clones may also be made of human organs for **transplants**.
- Cloning mammals has raised concern amongst the general public and some scientists.

Some arguments for and against cloning

For	Against
Saving endangered species	Embryos are destroyed
Providing organs for transplants	Made in human's mould, not God's
An end to genetic and other diseases	Prevents natural evolution
Childless couples could have children	Clones may develop abnormalities and defects
Exact copies of transgenic organisms can be produced	

Designer babies

- It may soon be possible to choose the physical characteristics and personality of our children.
- Embryos that carry the desired genes can be selected and implanted into a female.
- It may also be possible to alter the genes in human embryos to give the offspring particular features.

Definitions

- **Transgenic organism** an organism containing genes taken from another species
- **Transplant** an organism that was donated by one organism and inserted into the body of another, e.g. kidney transplant

Look on the CD for more exam practice questions

❶ Examiner's tips

- There are many social, ethical and legal issues surrounding this topic and you must be aware of both the advantages and disadvantages in producing transgenic organisms, clones and designer babies. Just stating it's wrong or right is not enough to gain marks in an examination.

Can you answer these questions?

1. Name two substances that can be produced by transgenic organisms.
2. Give two advantages and two disadvantages of cloning other than those stated in the table.
3. Explain what is meant by the terms 'designer milk' and 'designer babies'.

Did you know?

- A spider gene has been introduced into a goat, which now produces silk in its milk for use in bullet-proof vests and other materials.

Electrical and chemical signals

1 The nervous system

Have you ever wondered?

How does my brain tell my body what to do?

Key facts

- The **brain** and spinal cord make up the **central nervous system (CNS).**
- Sensory neurones carry electrical impulses from receptors in the sense organs to the CNS.
- The CNS sorts out (coordinates) the incoming information and sends out impulses to effector organs along motor neurones.
- Impulses travel very quickly through our neurones and allow us to respond very rapidly to changes going on inside and around us.

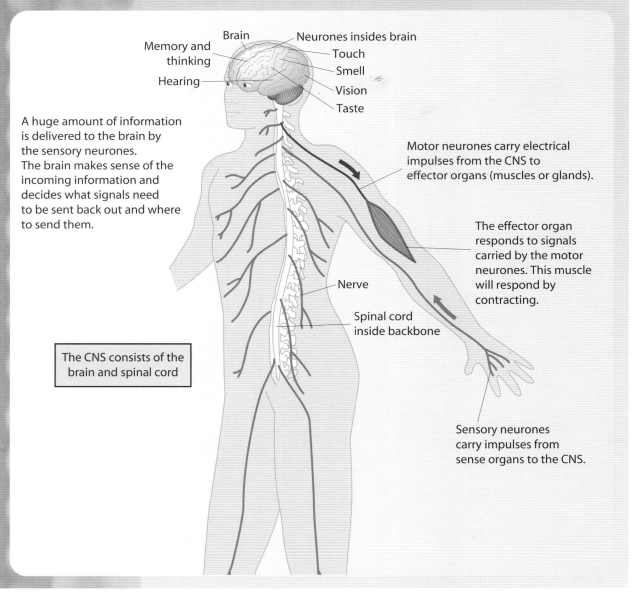

Brain
Memory and thinking
Hearing
Neurones insides brain
Touch
Smell
Vision
Taste

A huge amount of information is delivered to the brain by the sensory neurones.
The brain makes sense of the incoming information and decides what signals need to be sent back out and where to send them.

Motor neurones carry electrical impulses from the CNS to effector organs (muscles or glands).

The effector organ responds to signals carried by the motor neurones. This muscle will respond by contracting.

Nerve

Spinal cord inside backbone

The CNS consists of the brain and spinal cord

Sensory neurones carry impulses from sense organs to the CNS.

● Certain disorders affect how the brain works.

Disorder	How it affects the brain
Stroke	A blood clot may prevent oxygen getting to brain cells. Cells become damaged or could die.
Tumours	These are cells which divide uncontrollably. They put pressure on delicate brain cells and can interfere with electrical impulses, causing unconsciousness and seizures.
Parkinson's disease	Communication between neurones is disrupted. Muscle cells don't receive clear signals and they produce jerky movements.
Grand mal **epilepsy**	Random signals are sent out by the brain to the body which may cause unconsciousness and uncontrolled movement (seizures).

Definitions

● **Central nervous system** (CNS) the brain and spinal cord
● **Brain** an organ that coordinates the actions of the body
● **Epilepsy** too much electrical activity in the brain – causes a seizure

Look on the CD for more exam practice questions

❶ Examiner's tips

● Students often state that nerves (neurones) send 'messages'. Although this is true in a sense, it does not distinguish between electrical or chemical messages so you would not gain any marks for this answer in an exam. Always state that neurones transmit or carry *impulses* or *electrical signals*.

Can you answer these questions?

1. What is the CNS made up of?
2. What is the role of the brain in the CNS?
3. Name the type of neurone that carries impulses to the brain.
4. Explain how grand mal epilepsy affects the brain.

Did you know?

● A typical computer would have to be a million times more powerful to perform like the human brain.

② Sense organs

Have you ever wondered?

Why we can get an electric shock from electricity in our home but not from the electricity in our body.

Key facts

● **Sense organs** contain special cells called **receptors**.
● Receptors are sensitive to any changes that occur inside or outside of our body.

- These changes are known as **stimuli** as they *stimulate* the receptors in our sense organs.
- Inside the sense organ, the receptors convert the stimulus into electrical impulses which travel along sensory neurones to the CNS.

The diagram shows the pathway that the impulses take when a receptor is stimulated.

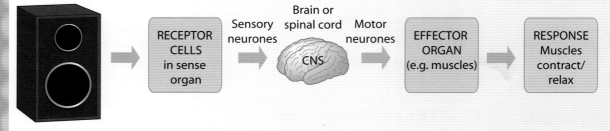

- Different sense organs are sensitive to different stimuli.

Sense organ	Stimulus
Eye	light intensity
Ears	pitch and loudness of sound, balance
Skin	pain, pressure, temperature
Tongue	sweet, bitter, salty and sour tastes
Nose	different odours

Definitions

- **Stimulus** (plural stimuli) something sound, heat, light – you react to
- **Receptors** a cell that detects a stimulus, e.g. light, sound, heat
- **Sense organ** an organ that has receptors to detect stimuli

Look on the CD for more exam practice questions

❶ Examiner's tips

- This topic crops up often in exams. Make sure that you are familiar with the route that electrical impulses take once a receptor is stimulated. The pathway is the same whatever the sense organ – the only things that will differ are the stimuli and the response.

Can you answer these questions?

1 What is a receptor?
2 Draw a flow diagram, like the one above, to show the pathway that impulses take when we touch a hot object.
3 What type of external change would stimulate receptors in our eyes?

Did you know?

- Mosquitoes like smelly feet. They are attracted to humans by their body odour.

③ The eye

Have you ever wondered?

Why our eye 'sees' upside down images?

Key facts

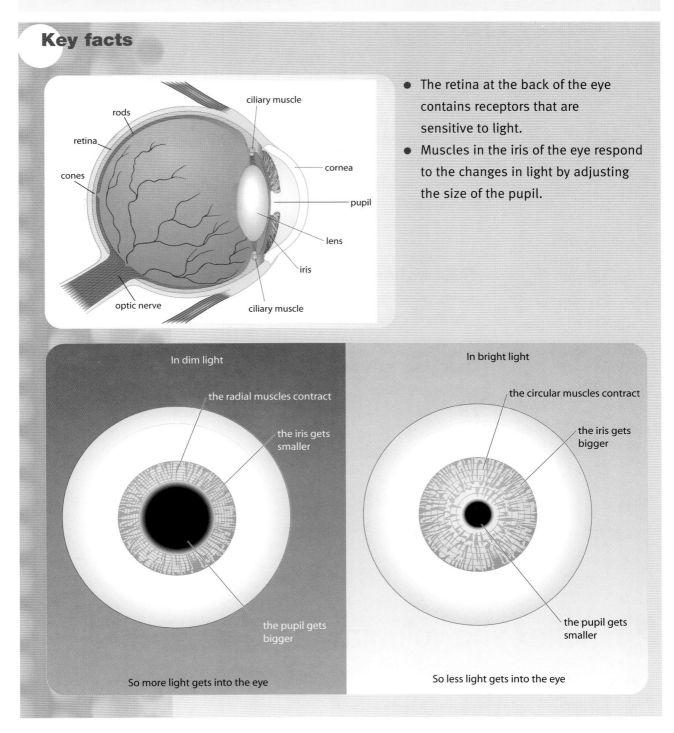

- The retina at the back of the eye contains receptors that are sensitive to light.
- Muscles in the iris of the eye respond to the changes in light by adjusting the size of the pupil.

ciliary muscle
rods
retina
cones
cornea
pupil
lens
iris
optic nerve
ciliary muscle

In dim light

the radial muscles contract

the iris gets smaller

the pupil gets bigger

So more light gets into the eye

In bright light

the circular muscles contract

the iris gets bigger

the pupil gets smaller

So less light gets into the eye

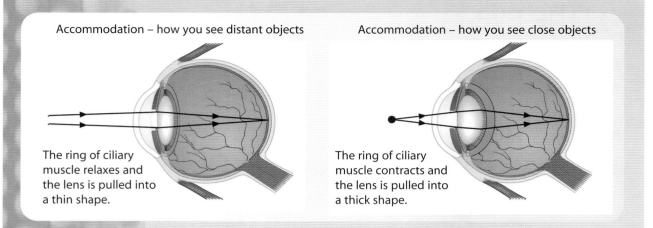

Accommodation – how you see distant objects

Accommodation – how you see close objects

The ring of ciliary muscle relaxes and the lens is pulled into a thin shape.

The ring of ciliary muscle contracts and the lens is pulled into a thick shape.

The stimulus (light) stimulates the receptor cells in the eye to make the muscles of the iris contract or relax.

- The ciliary muscles also respond by changing the shape of the lens to ensure that light is focussed on the retina.
- This is known as accommodation.

Definitions

- **Iris reflex** a reflex action that controls how much light enters the eye

Look on the CD for more exam practice questions

❶ Examiner's tips

- Don't mention *size*! The size of the lens does not change when we look at near or distant objects. It is only the *shape* that matters in this case.

Can you answer these questions?

1 What is accommodation?
2 What part of the eye contains light-sensitive receptors?
3 What is the iris reflex?
4 What type of neurone carries impulses to the muscles in the eye?

Did you know?

- The eye of a dragonfly contains 30 000 lenses.

④ Reflex and voluntary actions

Have you ever wondered?

Whether our brain stores images in the same way as a digital camera?

Key facts

- The iris reflex, accommodation in the eye and ducking are examples of **reflex** actions.
- Reflex actions are **involuntary** – they happen automatically and we have no control over them.

- Reflex actions are very fast responses to certain stimuli and they help to protect our body from potential harm. The pathway of a reflex action is shown by a **reflex arc**.

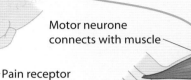

Sensory neurone

Synapse. This is the gap between two neurones. Chemicals travel across the synapse and trigger an impulse in the next neurone.

Nerve impulse from pain receptor

Motor neurone

Spinal cord

A relay neurone connects a sensory neurone to a motor neurone

Motor neurone connects with muscle

Pain receptor

Muscle

- **Voluntary responses** are actions that we can control.
- Unlike reflex actions, the brain is involved in a voluntary response as we need to think about carrying out the action, like moving our arm or kicking a ball.

Reaction times

- The time between detecting a stimulus e.g. hearing a sound and carrying out a response e.g. turning your head is called your **reaction time.**
- You can measure your reaction time by catching a falling ruler but there are more accurate electronic methods.

Definitions

- **Reflex** automatic response to a stimulus – cannot be controlled
- **Voluntary response** non-automatic response to a stimulus – can be controlled
- **Reaction time** the time your body takes to react to a stimulus

🛈 Examiner's tips

- Exam questions on reflex actions may require you to apply your knowledge to a situation that you are not familiar with. Just remember that the pathway is always the same – the basic principles will always apply regardless of the context of the question.

Can you answer these questions?

1 Why are reflex actions important?

2 Give two examples of reflex actions.

3. What organs does the ducking reflex protect?

4. Why are reflex actions quicker than voluntary actions?

Look on the CD for more exam practice questions

Did you know?

- Nerve impulses can travel over 320 km per hour which is 3 million times faster than the speed of electricity through a wire.

⑤ Chemical messengers

Have you ever wondered?

How do my hormones 'know' where to go?

Key facts

● Our blood is mainly liquid containing different types of cells.

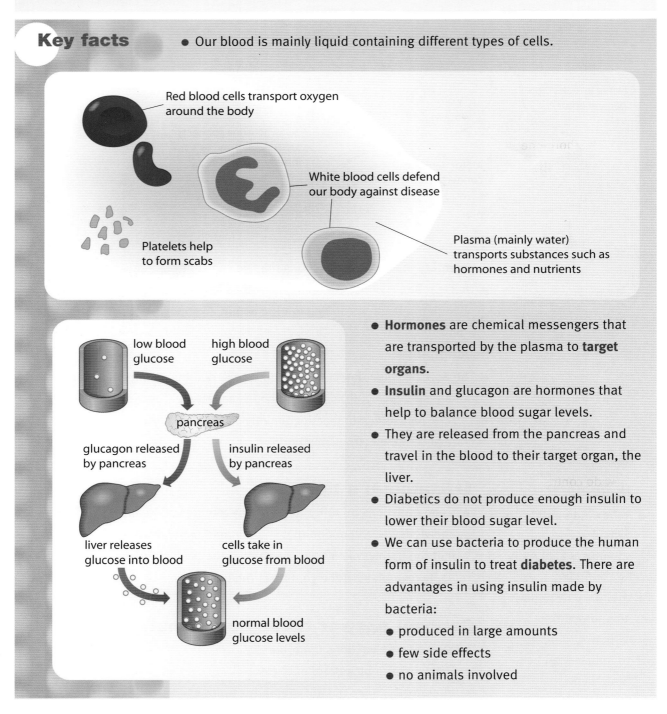

Red blood cells transport oxygen around the body

White blood cells defend our body against disease

Platelets help to form scabs

Plasma (mainly water) transports substances such as hormones and nutrients

low blood glucose

high blood glucose

pancreas

glucagon released by pancreas

insulin released by pancreas

liver releases glucose into blood

cells take in glucose from blood

normal blood glucose levels

● **Hormones** are chemical messengers that are transported by the plasma to **target organs**.

● **Insulin** and glucagon are hormones that help to balance blood sugar levels.

● They are released from the pancreas and travel in the blood to their target organ, the liver.

● Diabetics do not produce enough insulin to lower their blood sugar level.

● We can use bacteria to produce the human form of insulin to treat **diabetes**. There are advantages in using insulin made by bacteria:

  ● produced in large amounts

  ● few side effects

  ● no animals involved

Definitions

- **Hormone** a chemical message that coordinates the body –produced by a gland
- **Target organ** the organ on which a particular hormone works
- **Insulin** a hormone that controls the level of sugar in the blood – made in the pancreas
- **Diabetes** a disease caused by uncontrolled sugar levels in the blood

❶ Examiner's tips

- Hormones and nerve impulses are both messenger systems. Make sure you know some of the differences between the two.

Hormones	Nerve impulses
chemical	electrical
travel in blood	travel through neurones
slow response	rapid response

Can you answer these questions?

1 What is a hormone?
2 Name the hormone that lowers blood sugar levels.
3 What part of the blood transports hormones?
4 What are the advantages of using insulin from bacteria to treat diabetes?

Look on the CD for more exam practice questions

Did you know?

- Earthworms and insects have green blood, whereas lobsters and crabs have blue blood as their blood contains copper instead of iron.

⑥ Fertility and infertility

Have you ever wondered?

How do contraceptive pills work?

Key facts

Menstruation is:
- a 28-day cycle that prepares the uterus to receive a fertilised egg
- controlled by two main sex hormones, oestrogen and progesterone.

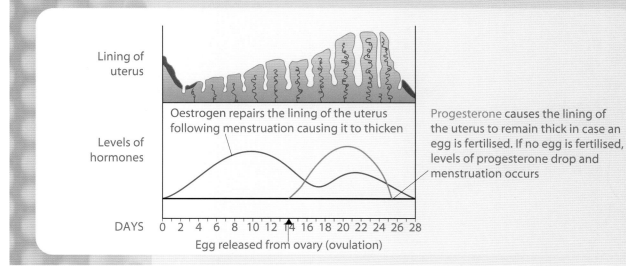

Lining of uterus

Levels of hormones

Oestrogen repairs the lining of the uterus following menstruation causing it to thicken

Progesterone causes the lining of the uterus to remain thick in case an egg is fertilised. If no egg is fertilised, levels of progesterone drop and menstruation occurs

DAYS 0 2 4 6 8 10 12 14 16 18 20 22 24 26 28

Egg released from ovary (ovulation)

- Sex hormones are used in contraceptive pills to prevent pregnancy.
- Some **contraceptives,** such as the pill, prevent the ovary from releasing an egg so no fertilisation can take place.
- Some sex hormones can be used to treat **infertility**.
- These hormones cause a greater number of eggs to be produced.
- The eggs are collected and mixed with sperm outside the body.
- A fertilised egg is placed back into the mother.
- This treatment is known as *in-vitro* **fertilisation** (IVF).

Definitions

- **Infertility** when a couple cannot have children, e.g. blocked oviduct, insufficient sperm
- *in-vitro* **fertilisation (IVF)** the fertilisation of a human egg outside the body
- **Contraception** a way to prevent pregnancy

Look on the CD for more exam practice questions

❶ Examiner's tips

- Make sure you know the arguments for and against IVF. Here are just a couple.

For	Against
Allows childless couples to have children.	Some couples may be too old to be parents.

Can you answer these questions?

1 What hormones control menstruation?

2 What is a contraceptive pill?

3 What is IVF?

4 Give one advantage and one disadvantage of IVF.

Did you know?

- IVF increases the chance of multiple births. In America, a woman gave birth to 8 babies (octuplets) but not all survived.

Use, misuse and abuse

1 Tuberculosis (TB)

Have you ever wondered?

Why is TB in the news again?

Key facts

- Tuberculosis is caused by a bacterium.
- Tuberculosis is spread through the air.
- You catch TB by spending a lot of time with an infected person.
- People with AIDS often also get TB.
- Some people do not finish their course of antibiotics.
- If this happens, some bacteria are not killed but can reproduce. Their offspring are harder to kill. This results in resistant bacteria.
- TB bacteria can hide in a human body as spores and can survive for many years and then cause **disease**.

Definitions

- **Disease** caused by microorganisms – it makes you ill

Look on the CD for more exam practice questions

❶ Examiner's tips

- People who are poor, live in overcrowded conditions and do not eat properly are at most at risk from catching disease.
- Drugs are used to treat disease. A new drug costs millions of pounds to develop.
- Failure to complete a course of antibiotics can result in resistant bacteria.
- TB is increasing in some parts of London.

Can you answer these questions?

1. What is an antibiotic?
2. How can some bacteria become resistant to an antibiotic?
3. Why do so many AIDS sufferers die from TB?
4. Why does a new drug cost so much to develop?

Did you know?

- 50% of all AIDS sufferers will catch TB, and it will kill half of them.

2 Microorganisms and disease

Have you ever wondered?
What is the difference between an infection and a disease?

Key facts

- Microorganisms like TB are pathogens. They cause disease.
- Most microorganisms do not cause disease.
- Microorganisms can be transmitted from one person to another in many ways.
- Your body has three lines of defence against microorganisms.
- The first line of defence uses chemical and physical barriers.
- The second line of defence uses white blood cells which ingest microorganisms.
- The third line of defence uses antibodies to destroy microorganisms.

Definitions
- **Inflammation** a response to infection – causes swelling and redness
- **Viral infection** an infection caused by a virus – antibiotics cannot cure it

❶ Examiner's tips
- Know the difference between direct and indirect contact.
- **Inflammation** is part of the body's response to infection.
- Disease is different from infection.
- Antibodies are part of the body's defence system.
- Antigens are the part of a bacterium that antibodies use to recognise and destroy the bacterium.

Can you answer these questions?
1. How are microorganisms removed from your breathing system?
2. Why is the second line of defence against microorganisms called the non-specific immune response?
3. Explain how vaccination gives you immunity.
4. What is the DOTS treatment and why is it very successful against TB?

Look on the CD for more exam practice questions

Did you know?
- Smallpox was the first disease to be completely eradicated by vaccination.

③ Types of drugs

Have you ever wondered?

How do different drugs affect people differently?

Key facts

- Many drugs affect the way the nervous system works.
- Some drugs, like caffeine, are stimulants and increase the rate at which impulses travel along the nervous system.
- Stimulants can increase alertness and improve **reaction time**.
- Alcohol and barbiturates are **sedatives** and slow down impulses in the nervous system.
- Addiction is when someone is dependent on a drug and can't do without it.
- The body can get used to a drug and then needs increasing amounts to get the same effect.
- Some drugs are used for pain relief.

Definitions

- **Sedative** a type of drug – can slow the nervous system down
- **Reaction time** the time that your body takes to react to a stimulus

❶ Examiner's tips

- You need to know how alcohol affects the body and how you behave, and why it is dangerous to drink and drive.
- Think about why sports people take caffeine-based drinks to help them, but are banned from using performance enhancing drugs.
- Drugs:
  - can be stimulants or sedatives
  - can cause addiction
  - can be used for pain relief.

Look on the CD for more exam practice questions

Can you answer these questions?

1. Name two sedative drugs
2. Explain how stimulants affect your reaction time.
3. Explain why doctors are very careful when prescribing barbiturates for epilepsy.
4. What effect do drugs have on synapses?

Did you know?

Caffeine is a drug found in tea, coffee, chocolate and some cola drinks.

④ Pain relief

Have you ever wondered?

Why are some drugs considered good for your body and others bad?

Key facts

- Paracetamol is the most widely used drug for pain relief.
- Too much paracetamol can severely damage the liver.
- Taking more than the prescribed dose is called an overdose.
- Morphine is a very strong painkiller, used for people in severe pain.
- Morphine use can lead to addiction.
- Heroin is made from morphine, but is even more addictive.
- Heroin is injected. Sharing needles increases the risk of contacting AIDS.

Definitions

- **Opiate** a drug from the opium family, e.g. opium, heroin

❶ Examiner's tips

- Pain relief drugs have disadvantages as well as advantages.
- Some pain relief drugs can be addictive.
- The advantages of cannabis as a pain relief drug are outweighed by the dangers of smoking it.

Look on the CD for more exam practice questions

Can you answer these questions?

1. Name three painkillers.
2. How do painkillers affect the nervous system?
3. Explain why morphine and heroin can be very dangerous.
4. What can be the result of taking too much paracetamol?

Did you know?

- Morphine comes from the opium poppy.

⑤ Drug misuse and abuse

Have you ever wondered?

Why are the uses of some substances controlled by law?

Key facts

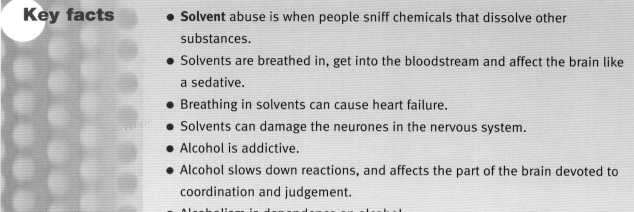

- **Solvent** abuse is when people sniff chemicals that dissolve other substances.
- Solvents are breathed in, get into the bloodstream and affect the brain like a sedative.
- Breathing in solvents can cause heart failure.
- Solvents can damage the neurones in the nervous system.
- Alcohol is addictive.
- Alcohol slows down reactions, and affects the part of the brain devoted to coordination and judgement.
- Alcoholism is dependence on alcohol.

Definitions

- **Solvent** a chemical used to dissolve other substances, e.g. aerosols, thinners (may be abused by sniffing)

❶ Examiner's tips

- Alcohol slows down reactions, increasing reaction time.
- Solvents have lots of different effects – do you know them all?
- Alcohol affects:
  - the brain
  - the liver
  - behaviour.

Look on the CD for more exam practice questions

Can you answer these questions?

1. Describe three effects of solvent abuse.
2. How do solvents affect neurones?
3. What is alcoholism?
4. Why is it dangerous to drink and drive?

Did you know?

- In 2004 there were more than 17 000 drink-drive casualties.

(6) Tobacco

Key facts

- Cigarettes contain nicotine.
- Nicotine is a very addictive drug. It causes craving and is incredibly difficult to give up.
- Nicotine increases blood pressure and can cause heart disease.
- Burning tobacco produces carbon monoxide. This reduces the ability of the blood to carry oxygen.
- Nicotine and carbon monoxide together make the blood clot more easily, increasing the risk of a heart attack.
- Tar from burning tobacco contains chemicals that cause cancer.
- If the air passages to the lungs become infected this can cause bronchitis.

Definitions

- **Gaseous exchange** the way your lungs take in oxygen and expel carbon dioxide
- **Cilia** tiny hair-like structures that clear mucus from lungs and windpipe

❶ Examiner's tips

- Cigarettes can affect the body in many ways. Do you know them all?
- Nicotine is addictive and causes craving. It is hard to give up.
- Look at graphs of data showing trends in smoking . Can you explain the shape of the graph?

How the percentage of adult smokers has changed since 1974

	1974	1978	1982	1986	1990	1994	1996	1998	2000	2003
Men	51	45	38	35	31	28	29	28	29	28
Women	41	37	33	31	29	26	28	26	25	24
All	45	40	35	33	30	27	28	27	27	26

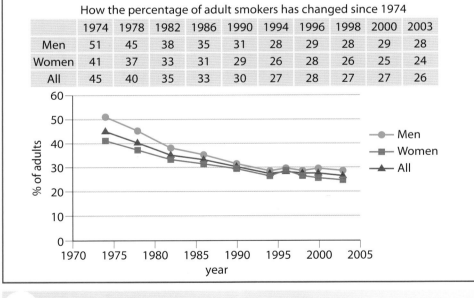

Can you answer these questions?

1. Name three substances from cigarettes that can affect the body.
2. How does smoking affect your heart?
3. Why do some long-term smokers have to have extra oxygen?
4. What are the effects of bronchitis?

Did you know?

- More than 1600 people die each week in England from smoking-related illness.

Look on the CD for more exam practice questions

Patterns in properties

1 The periodic table

Have you ever wondered?

Is the periodic table a map of what you're made from?

Group 1	2										3	4	5	6	7	0
Period 1																4 He helium 2
7 Li lithium 3	9 Be beryllium 4										11 B boron 5	12 C carbon 6	14 N nitrogen 7	16 O oxygen 8	19 F fluorine 9	20 Ne neon 10
23 Na sodium 11	24 Mg magnesium 12										27 Al aluminium 13	28 Si silicon 14	31 P phosphorus 15	32 S sulphur 16	35.5 Cl chlorine 17	40 Ar argon 18
39 K potassium 19	40 Ca calcium 20	45 Sc scandium 21	48 Ti titanium 22	51 V vanadium 23	52 Cr chromium 24	55 Mn manganese 25	56 Fe iron 26	59 Co cobalt 27	59 Ni nickel 28	63.5 Cu copper 29	65 Zn zinc 30	70 Ga gallium 31	73 Ge germanium 32	75 As arsenic 33	79 Se selenium 34	80 Br bromine 35	84 Kr krypton 36
85 Rb rubidium 37	88 Sr strontium 38	89 Y yttrium 39	91 Zr zirconium 40	93 Nb niobium 41	96 Mo molybdenum 42	(98) Tc technetium 43	101 Ru ruthenium 44	103 Rh rhodium 45	106 Pd palladium 46	108 Ag silver 47	112 Cd cadmium 48	115 In indium 49	119 Sn tin 50	122 Sb antimony 51	128 Te tellurium 52	127 I iodine 53	131 Xe xenon 54
133 Cs caesium 55	137 Ba barium 56	139 La lanthanum 57	178 Hf hafnium 72	181 Ta tantalum 73	184 W tungsten 74	186 Re rhenium 75	190 Os osmium 76	192 Ir iridium 77	195 Pt platinum 78	197 Au gold 79	201 Hg mercury 80	204 Tl thallium 81	207 Pb lead 82	209 Bi bismuth 83	(209) Po polonium 84	(210) At astatine 85	(222) Rn radon 86
(223) Fr francium 87	(226) Ra radium 88	(227) Ac actinium 89	(261) Rf rutherfordium 104	(262) Db dubnium 105	(266) Sg seaborgium 106	(264) Bh bohrium 107	(277) Hs hassium 108	(268) Mt meitnerium 109	(271) Ds darmstadtium 110	(272) Rg roentgenium 111						

relative atomic mass ——— 1
H
hydrogen
atomic number ——— 1

Key facts

- The periodic table is an alphabet of chemistry.
- There are about 116 **elements**.
- Different combinations of elements result in millions of **compounds**.
- The elements are listed in order of increasing atomic number.
- The vertical columns are called groups.
- Elements with the same number of outer electrons appear in the same group.
- Elements in a group share characteristic properties.
- The horizontal rows are called periods.
- Each element has its own symbol.

❶ Examiner's tips

- Take care with upper and lower case letters when you use atomic symbols. Hf represents an atom of hafnium (a transition metal) but HF represents a molecule of hydrogen fluoride (a halogen compound). They are very different substances!

Definitions

- **Elements** substances that are made of only one type of atom
- **Compound** atoms of two or more different elements chemically joined to form a substance

Look on the CD for more exam practice questions

Can you answer these questions?

1. What are the symbols for
   (a) chlorine   (b) cobalt   (c) carbon   (d) calcium?
2. What are the names of these elements?
   a) Ne   (b) Na   (c) N   (d) Ni
3. What is the chemical difference between Co and CO?

Did you know?

- The only letter that does not appear on the periodic table is J.

② The atom

Have you ever wondered?

Can chemists turn cheap metals into gold?

Key facts

- All elements are made up of **atoms**.
- Atoms themselves are made from even smaller particles.
- There are three types of sub-atomic particle – electrons, protons and neutrons.
- The atomic number is equal to the number of protons.
- No two elements have the same atomic number.
- Uncombined atoms have no charge because the positively and negatively charged particles balance out.

Particle	Charge	Found in
Proton	positive	nucleus
Neutron	none	nucleus
Electron	negative	shells/ orbits surrounding nucleus

electron — proton — neutron

Definitions

- **Atoms** the smallest particles of an element that can exist.

Look on the CD for more exam practice questions

Can you answer these questions?

1. How many protons are present in the nucleus of these atoms?
   (a) calcium   (b) oxygen   (c) krypton
2. Name the elements with the atomic numbers
   (a) 79   (b) 82   (c) 47

Did you know?

- There are more molecules in a bucket full of water than there are buckets of water in the Atlantic Ocean.

③ Analysis and identification

Have you ever wondered?

How forensic scientists can identify traces of substances at a crime scene.

Key facts

- Some metals in compounds give characteristic colours when heated in a Bunsen flame. This means we can use a flame test to identify them.

Metal	Flame test colour
lithium	red
sodium	orange
potassium	lilac
calcium	yellow-red
barium	pale green
copper	green-blue

- Solutions of transition metals form coloured **precipitates** with sodium hydroxide solution. This is another way to identify these metals in their compounds.

Transition metal	Colour of precipitate
cobalt	blue (turns grey when left standing)
copper	pale blue
iron(II)	dark green (turns orange-brown when left standing)
iron(III)	orange-brown
manganese	pale brown
nickel	green
zinc	white

Definitions

- **Precipitate** solid formed when two solutions are mixed together

Look on the CD for more exam practice questions

❶ Examiner's tips

- These reactions and colours are tricky to learn from a book. The best way is to do practical work and see the reactions for yourself. In exams you will be presented with data and asked to draw conclusions. Take time to read the data carefully and make sure you use it in your answers.

Can you answer these questions?

1. Which metals would give these colours in the flame test?
   (a) red
   (b) pale green
   (c) orange
2. What metal is present in these coloured precipitates formed by reaction with sodium hydroxide solution?
   (a) green
   (b) pale blue
   (c) orange-brown
3. What colour precipitate would iron(II) give in sodium hydroxide solution?

Did you know?

- Forensic scientists use a spectrophotometer to show colours in very small or dilute samples of evidence. At these levels the human eye could not detect the colours.

④ Alkali metals

Have you ever wondered?

Which combination of substances make the most violent explosions?

Key facts

Alkali metals are all in group 1 in the periodic table. This group contains the most reactive metals. They have these properties:

- they are solid at room temperature
- they conduct heat and electricity
- they are soft enough to be cut with a knife
- they are metallic and shiny when first cut
- they all have low densities; lithium, sodium and potassium are so light they float on water
- they become more reactive as their atomic number increases down the group
- they tarnish quickly in air and need to be stored in oil
- they all react vigorously with water to form alkalis and hydrogen
- these reactions all produce heat; reacting potassium gets so hot that the hydrogen released catches fire.

Definitions

- **Diatomic molecule**
  two atoms joined
  by chemical bonds
  to make a molecule

Look on the CD for more exam practice questions

Can you answer these questions?

1. Write a word equation for the reaction between sodium and water.
2. Which fruit is represented by an atom of barium and two atoms of sodium? (Clue: look at the atomic symbols.)

Did you know?

- The reaction between caesium and water can be so vigorous that just two grams thrown into a bathtub of water can blow the bath in two.

⑤ Halogens

Have you ever wondered?

Why is chlorine so good at protecting you from other people's bugs in a swimming pool?

Key facts

The halogens are a reactive group of non-metals. They have these properties:

- their atoms all have 7 electrons in their outer shell
- they exist as molecules of two atoms joined together
- their melting points increase as the atomic number increases down the group
- they react with metals to form salts
- their reactivity decreases with increasing atomic number and as their size increases
- the more reactive halogens will displace the less reactive ones from their compounds
- they are all oxidising agents
- they are all toxic
- hydrogen halides react with water to form strong acids.

Element	Symbol	Colour
fluorine	F	pale yellow
chlorine	Cl	yellow/green
bromine	Br	orange/brown
iodine	I	shiny grey

❶ Examiner's tips

- The chemistry of the halogens is all about patterns and trends. Exam questions often rely on these patterns and ask for predictions.

Look on the CD for more exam practice questions

Can you answer these questions?

1. List the halogens in order of reactivity. Start with the least reactive.
2. Suggest what astatine might look like.

Did you know?

● Iodine can change from being a solid directly into a gas without going through a liquid stage. This is called subliming.

6 Noble gases

Have you ever wondered?

What chemicals are used in a laser light show?

Key facts

The noble gases are unreactive. They have these properties:
● they exist as single atoms
● they are all colourless and odourless
● all atoms of noble gases have 8 electrons in their outer shell, except helium atoms which have 2
● the gases become denser with increasing atomic number – argon, krypton and xenon are heavier than air
● electricity can make the noble gases glow and each gas glows with its own distinct colour
● noble gases are found in small quantities in the air and are obtained by **fractional distillation** of liquid air.

Definitions

● **Fractional distillation** distillation that gives a mixture of liquids (called fractions) instead of complete separation

Can you answer these questions?

1. Why are the noble gases so unreactive?
2. List the common uses of (a) helium   (b) argon.

Did you know?

● Sound is produced by vibration caused by the movement of air around our vocal chords. If you inhale helium gas, because it is less dense than air, it causes the vocal cords to vibrate faster, producing a higher-pitch which makes you sound like Donald Duck.

Look on the CD for more exam practice questions

Making changes

① Oxygen

Have you ever wondered?

How do you collect and test different gases?

Key facts

- Oxygen makes up 21% of the Earth's atmosphere by volume.
- Oxygen is colourless, odourless and only slightly soluble in water.
- **Oxidation** occurs when an element or compound gains oxygen in a reaction.
- Reduction occurs when a compound loses oxygen in a reaction.
- Combustion (burning) is a form of oxidation.
- Reactions between pure metals and oxygen form metal oxides.
- The test for oxygen is to place a glowing splint in a test tube and the splint will relight.

Definitions

- **Decomposition** a reaction where a substance breaks down to form two or more new substances
- **Oxidation** a reaction where oxygen is added to an element or compound

Look on the CD for more exam practice questions

❶ Examiner's tips

- Balanced equations must be written correctly. Take care with upper and lower case e.g. magnesium is written Mg, not mg or mG or MG.
- Oxygen is diatomic (its molecule consists of two atoms bonded together) so if you use oxygen in an equation it is always written O_2.
- State symbols should be included in balanced equations: solid (s), liquid (l), gas (g) and aqueous solution (aq) (dissolved in water).

Can you answer these questions?

1. Why does magnesium burn more brightly in oxygen than in air?
2. Write the word and balanced equations for the burning of magnesium in air (remember oxygen is diatomic).
3. How can you tell that the decomposition of hydrogen peroxide (H_2O_2) into water (H_2O) is a reduction reaction?

② Other important gases

Have you ever wondered?

How do the bubbles that make cakes so light get there?

Key facts

- Hydrogen is a colourless, odourless gas which is highly explosive when mixed with air and ignited.
- The test for hydrogen is to place a burning splint in a test tube of hydrogen and the gas ignites with a squeaky pop.
- Carbon dioxide is a colourless, odourless gas that is a product of respiration.
- The test for carbon dioxide is to bubble it through limewater which will turn milky.
- Carbon dioxide is sometimes produced when carbonates react and is a product of **combustion** when carbon-containing fuels are burnt.
- Gases can be collected by three methods, upward delivery (for gases lighter than air), downward delivery (for gases heavier than air) and over water (for gases which are insoluble or only slightly soluble in water).

Definitions

- **Combustion** a chemical reaction between fuel and oxygen – the burning process gives out heat

Look on the CD for more exam practice questions

❶ Examiner's tips

Follow these few simple rules and you should always be able to predict the products of reactions correctly.

- Whatever is on one side of an equation must also be on the other.
- Elementary hydrogen is diatomic and appears as H_2 in equations.
- Products of combustion reactions of fuels containing carbon and hydrogen only are carbon dioxide and water, if there is excess oxygen.

Can you answer these questions?

1. What are the products formed when methane (CH_4) burns in oxygen?
2. Limestone (calcium carbonate, $CaCO_3$) reacts with hydrochloric acid (HCl). What gas is formed in this reaction? Explain, in terms of the elements present in the compounds, why this gas is formed.
3. Describe the method of collecting hydrogen (which is lighter than air) and how you would test for it.

Did you know?

- The airship 'The Hindenberg' used hydrogen as a lighter than air gas. This undoubtedly contributed to the disastrous explosion that destroyed it on its maiden flight. Airships today use helium, which is an unreactive gas, so much safer.

3 Metals and reactivity

Have you ever wondered?

Did people always have metals?

Key facts

This is the reactivity series for metals.

most reactive	potassium	
	sodium	
	calcium	Extracted using electricity
	magnesium	
	aluminium	
	zinc	
	carbon	
	iron	
	tin	Extracted by reacting with carbon
	lead	
	copper	
	silver	
	gold	Found uncombined in Earth's crust
least reactive	platinum	

- Metals can be obtained from their ores which are rocks containing metal compounds.
- Metals can be put into a list according to their reactivity; the reactivity series.
- Those metals at the top of the list are the most reactive, those at the bottom the least reactive.
- The most reactive metals need most energy to extract them from their ores. For example, aluminium is extracted by electrolysis.
- Metals lower in the reactivity series can be extracted from their ores by heating with carbon.
- Metals with a high reactivity react with oxygen in the atmosphere, those with a low reactivity do not.
- If a metal is higher than another in the reactivity series, it can replace the other metal in a compound, displacing that metal in a **displacement** reaction. An example of this is:

magnesium + iron oxide → magnesium oxide + iron

Definitions

- Displacement reaction in which a metal higher in the reactivity series replaces one lower in the series in a compound

Look on the CD for more exam practice questions

Can you answer these questions?

1. What is the name of the gas formed when iron oxide is heated with carbon?
2. This reaction involves the loss of oxygen. What do we call this type of reaction?
3. Write out the names of the products of the reaction between calcium and copper oxide.

Did you know?

- Early man may have discovered iron when using stones containing the ore around a fire.

4 Metals and their salts

Have you ever wondered?

What's in a firework?

Key facts

- When metals react with acids, **salts** and hydrogen (H_2) are formed.
- Metals will react with **dilute** hydrochloric and sulphuric acids if they are above hydrogen in the reactivity series.
- Metal oxides react with acids to form metal salts and water.
- Metal hydroxides (alkalis) react with acids to form metal salts and water. This is called **neutralisation**.
- Metal carbonates form metal salts, water and carbon dioxide.
- If salts are insoluble in water they form a **precipitate**; most carbonates are insoluble.
- If salts are soluble in water they form a solution; all nitrates, most chlorides and most sulphates are soluble.
- The main acids are hydrochloric acid (HCl), nitric acid (HNO_3), and sulphuric acid (H_2SO_4).

Definitions

- **Dilute** less concentrated
- **Salt** substances formed in a neutralisation reaction – a compound where a metal has replaced the hydrogen in an acid
- **Precipitate** the insoluble solid formed in a precipitation reaction
- **Neutralisation** a reaction where an acid and an alkali (or other base) form a neutral solution

❶ Examiner's tips

- Make sure you know the names and chemical formulae of acids and their salts.
- Reactions with hydrochloric acid will form metal chlorides.
- Reactions with nitric acid will form metal nitrates.
- Reactions with sulphuric acid will form metal sulphates.
- You can recognise hydroxides by the OH group at the end of their formula, e.g. NaOH is the formula of sodium hydroxide.
- When balancing equations, remember that you cannot change the chemical formula of any element or compound, but you can show more than one substance on both sides of the arrow.

Look on the CD for more exam practice questions

Can you answer these questions?

1. Write the word and balanced equations for the reactions of these metals with sulphuric acid:
   (a) magnesium   (b) zinc
2. Write the balanced symbol equation for the reaction of sodium with chlorine (remember chlorine is diatomic).
3. Explain, in terms of atoms, why metal hydroxides form water, and not hydrogen, as one of the products when they react with water.

(5) Common compounds in food

Have you ever wondered?

Can you get cancer from eating too many food additives?

Key facts

- Thermal **decomposition** is the breakdown of a substance by the action of heat. This breakdown has uses in cooking where baking powder (sodium hydrogencarbonate) breaks down to form carbon dioxide, which causes cakes to rise.
- Acids are often used for flavouring food; **citric acid** and phosphoric acid are used in fizzy drinks such as cola.
- Carbohydrates contain carbon, hydrogen and oxygen. Sugar is a **carbohydrate**.
- Ethanoic acid, sometimes called acetic acid, is found in your kitchen cupboard and is commonly called vinegar.
- Artificial sweeteners are sometimes used to flavour foods instead of sugars but, in large amounts, some of these can have unpleasant effects such as diarrhoea.

Definitions

- **Hydration** a reaction where water is added
- **Dehydration** a reaction where water is removed
- **Citric acid** an acid found in oranges and lemons
- **Carbohydrate** compound that contains carbon, hydrogen and oxygen only

Look on the CD for more exam practice questions

⊕Examiner's tips

- Ensure you know the difference between natural and artificial (manufactured) substances; artificial substances have many advantages but can also cause problems.
- There are often reports in newspapers about problems with certain foods. It is important to understand that a correlation between a food and a problem does not necessarily mean that the food is the cause of the problem. Many other factors may be involved.

Can you answer these questions?

1. What may be a problem associated with adding phosphoric acid to fizzy drinks?
2. Write a word equation for the thermal decomposition of sodium hydrogencarbonate. Two of the products are sodium carbonate and water.
3. What type of reaction is the breakdown of copper carbonate using heat?
4. Why is the production of carbon dioxide gas beneficial during cake making?

Did you know?

- Sodium chloride is simply the chemical name for the salt you put on your food.

6 Chemicals for cleaning

Have you ever wondered?
Are the cleaning substances we use in our homes harmless?

Key facts
- Alkalis such as sodium hydroxide are used in some cleaning products.
- Ammonia is a strong-smelling substance which is used in the preparation of fertilisers and bleaches.
- Ammonia is produced by heating ammonium chloride with calcium hydroxide.
- Ammonia is soluble in water and must be collected by upward delivery. You can test for ammonia using damp red litmus paper which turns blue.
- Acids are also used in some cleaning products.

⊕ Examiner's tips
- Cleaning products often have hazard labels. Familiarise yourself with these as it is important that you can identify potential dangers in household and laboratory products. Here are some common hazard symbols.

Toxic	**Highly flammable**	**Corrosive**

These substances can cause death.	These substances catch fire easily.	These substances destroy living tissue.

Harmful	**Oxidising**	**Irritant**
Similar to toxic substances but less dangerous.	These provide oxygen for other substances to burn.	Not corrosive but may cause blisters or red skin.

Look on the CD for more exam practice questions

Can you answer these questions?
1. What are the other products formed during the production of ammonia from ammonium chloride and calcium hydroxide?
2. If a household cleaner can blister the skin on contact, which hazard symbol should appear on the label?
3. Why are oxidising agents sometimes used in the making of fireworks?

There's one Earth

1 Early Earth to the present day

Have you ever wondered?

Why do some scientists need to work in exotic locations like Hawaii and Antarctica?

Key facts

- The Earth was formed approximately 4.6 billion years ago.
- Volcanic activity in the early Earth released the gases carbon dioxide and water vapour into the atmosphere in large quantities. Small amounts of nitrogen, hydrogen and carbon monoxide were also released.
- The water vapour condensed to form the oceans of the Earth as the temperature of the Earth steadily dropped.
- Primitive green plants grew by the oceans and photosynthesised using carbon dioxide and water to produce glucose and oxygen. This caused the atmospheric oxygen levels to rise.
- The present composition by volume of the atmosphere is 78% nitrogen, 21% oxygen and 1% other gases, of which 0.035% is carbon dioxide.
- Air can be separated into its individual gases using **fractional distillation** but the air must be liquefied first then gradually allowed to warm. Nitrogen is given off at the top of the fractionating column.

Definitions

- **Fractional distillation** distillation that gives a mixture of liquids (called fractions) instead of complete separation

Can you answer these questions?

1. What gas made up most of the Earth's atmosphere 4 billion years ago?
2. What gas do plants need in order to photosynthesise?
3. How can you separate the gases in the air?
4. What caused the carbon dioxide levels in the atmosphere to be reduced over the past 4 billion years?

Did you know?

- Some of the oxygen in the Earth's atmosphere is converted into ozone (O_3) which protects the Earth from harmful ultraviolet rays.
- The Earth is 4.6 billion years old and 1 billion is equivalent to 1 000 000 000 therefore the Earth is 4 600 000 000 years old.

Look on the CD for more exam practice questions

② Global warming

Have you ever wondered?

Will the UK freeze over in one day as in the film *The day after tomorrow?*

Key facts

- The global temperature of the Earth has fluctuated from the average by between −8 °C and +4 °C due to the Earth going through various mini ice ages over the past 5000 years.
- Over the past 200 years, since the start of industrialisation, average global temperatures have risen by 1.5 °C, at the same time as carbon dioxide levels have risen.
- The greenhouse effect causes heat to be reflected back to the Earth's surface and not released into space, making the temperature of the Earth rise.
- **Global warming** is monitored by measuring sea levels and the size of polar ice sheets, in addition to the salinity (saltiness) of the ocean.
- We can reduce carbon dioxide emissions by reducing the use of energy produced from **fossil fuels** or investing in alternative energy resources.

Definitions

- **Fossil fuels** fuels e.g. coal, oil, gas that formed millions of years ago from the remains of plants and tiny animals
- **Global warming** a rise in the average temperature of the atmosphere and the Earth's surface

Look on the CD for more exam practice questions

❶ Examiner's tips

- All the methods of combating global warming are based on the precautionary principle which is, if we are not completely certain of the effects of something, we should act to prevent potential problems.
- Many factors may cause the greenhouse effect and global warming. These include:
  - burning fossil fuels, including petrol in cars, lorries, aeroplanes, power stations, etc
  - an increase in cattle and rice farming
  - deforestation.

Can you answer these questions?

1. What is the precautionary principle and how can you apply it to carbon dioxide levels in the atmosphere?
2. Name two methods of monitoring global warming.
3. Describe how deforestation can lead to a rise in carbon dioxide levels in the atmosphere.

Did you know?

- The Kyoto protocol is an agreement made between countries to cut greenhouse gas emissions by 5.2% by 2012, but the USA has still not signed up to the protocol.

3 Crude oil and fuels

Have you ever wondered?

As oil stocks run out, will petrol eventually cost as much as gold?

Key facts

- Crude oil is a mixture of hydrocarbon molecules. These must be separated into simpler mixtures to produce useful products such as petrol, kerosene and naptha. Hydrocarbons only contain hydrogen and carbon.

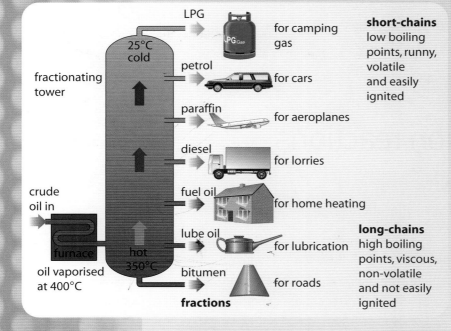

- Crude oil can be separated into fractions using a fractionating column like this.

- The viscosity of a liquid is a measure of how thick or runny a liquid is. Long chain hydrocarbons make oils with high viscosity, these are separated out towards the bottom of the fractionating column.

- There are alternatives to petrol, such as biofuel, which is made from living things. Hydrogen is considered to be an alternative fuel to petrol for cars but new technology will be needed to enable this to happen.

Definitions

- **Biofuel** fuel made from living things, e.g. ethanol from sugar beet, biodiesel from rapeseed oil
- **Fractionating column** a tall tower for industrial fractional distillation – the different fractions condense at different levels in the tower
- **Hydrocarbon** a compound that contains carbon and hydrogen atoms only
- **Crude oil** a natural mixture of hydrocarbons – a fossil fuel
- **Viscosity** how thick or runny a liquid is: high viscosity = thick, low viscosity = runny

❶ Examiner's tips

- Remember hydrocarbons contain *only* hydrogen and carbon.
- Questions about alternative fuels are popular in exams, so ensure you are familiar with the advantages and disadvantages of biofuel and hydrogen over petrol.

Look on the CD for more exam practice questions

Can you answer these questions?

1. Why is biofuel (made from living plants) considered to be carbon dioxide neutral?
2. Describe how the average length of carbon chains in a fractionating column changes as you move up the tower. How does their viscosity vary?
3. How many of these are formulae for hydrocarbon molecules?
C_2H_4, H_2SO_4, HCl, C_5H_{10}

Did you know?

- Many medicines and plastics are made from a fraction of crude oil called naptha.

4 Combustion reactions

Have you ever wondered?

How carbon monoxide can cause suffocation?

Key facts

- When hydrocarbon fuels burn they give out heat energy (an exothermic reaction). The more completely they burn, the more energy they give out.
- The equation for the complete combustion of methane in the presence of excess oxygen is:
$CH_4 + 2O_2 \rightarrow CO_2 + 2H_2O$
- Incomplete combustion occurs when there is insufficient oxygen available. During incomplete combustion, carbon and/or carbon monoxide may be formed.
- Carbon monoxide is a toxic gas which forms bonds with red blood cells when it enters your blood stream. This stops oxygen being carried around the body.
- The equations for the incomplete combustion of methane are:
$2CH_4 + 3O_2 \rightarrow 2CO + 4H_2O$
and
$CH_4 + O_2 \rightarrow C + 2H_2O$

Definitions

- **Combustion** a chemical reaction between fuel and oxygen – the burning process gives out heat

🛈 Examiner's tips

- If a question states that a fuel burns in air, this is referring to the use of oxygen in the air. You do not need to worry about the other components in the air as they are not involved in the burning process.
- Fossil fuels produce carbon dioxide when burnt. Carbon dioxide is a major contributor to global warming.

Look on the CD for more exam practice questions

Can you answer these questions?

1. Write a balanced equation for the complete combustion of ethane (C_2H_6) in air.
2. Describe the problems associated with breathing in carbon monoxide.
3. Fire fighters use various methods to put out fires. What gas present in the air are they trying to prevent reaching the fire, and why?

Did you know?

- Faulty gas appliances can cause carbon monoxide to be produced, so if there are gas appliances in your house, ensure they are checked regularly.

⑤ Recycling

Have you ever wondered?

Why do we recycle so little of our rubbish in this country?

Key facts

- The **recycling** of waste can save energy. It takes 20% less energy to recycle glass than it does to make it originally.
- Recycling reduces the amount of rubbish which has to be put in landfill sites or incinerated (burnt).
- Plastic waste is particularly difficult to dispose of as it releases toxic gases if burnt or takes thousands of years to decompose. Only biodegradable plastics break down naturally in the environment but new plastics are now being made which are recyclable.
- Metals such as aluminium are extracted from their ore using electrolysis. This is an expensive process, therefore recycling aluminium is cheaper and uses less energy.
- **Desalination** plants, used in hot countries, make sea water usable by removing the salt from it. This can be done using fossil fuel energy or solar power to heat up and evaporate the water.

❶Examiner's tips

- A key point in this section is the idea of **sustainable** development. This means allowing for continued development without compromising future generations. This relates not only to recycling but also to finding alternative fuels which have less impact on the environment.

Definitions

- **Desalination** changing salt water into fresh water by separating water from the dissolved salts
- **Recycle** reusing the same substance many times
- **Sustainability** being able to keep doing the same thing over and over again without harming the environment

Can you answer these questions?

1. What are the benefits of recycling glass, both economically and in terms of sustainable development?
2. What are the benefits of solar powered desalination plants as opposed to those powered by fossil fuels?
3. Why is sustainable development important for future generations?

Did you know?

- By recycling we reduce pollution as well as ensuring that the planet's resources are not depleted.

Look on the CD for more exam practice questions

⑥ The use of sea water

Have you ever wondered?

How we obtain gases from sea water?

Key facts

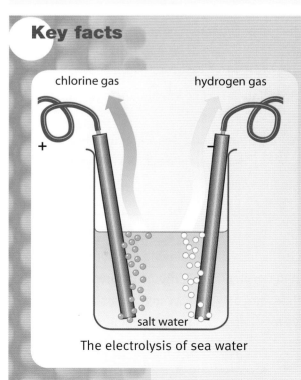

The electrolysis of sea water

- Sea water is the source of the gases we need to produce many everyday chemicals, such as chlorine for making bleaches.
- Sodium hydroxide can also be obtained from sea water and is used in the production of soaps and detergents.
- Hydrogen for making rocket fuel can also be produced from sea water
- In electrolysis, the positive hydrogen ions are attracted to the negative cathode and the negative chloride ions are attracted to the positive anode. Sodium hydroxide is also formed, in solution, as a result of the electrolysis of sea water.
- Sodium metal is not produced by the electrolysis of sea water but by the electrolysis of molten sodium chloride.

Definitions

- **Anode** positive electrode
- **Cathode** negative electrode

Look on the CD for more exam practice questions

❶ Examiner's tips

- Learn the word equation for the electrolysis of sea water:

  sodium chloride$_{(s)}$ + water$_{(l)}$ → chlorine$_{(g)}$ + hydrogen$_{(g)}$ + sodium hydroxide$_{(aq)}$

- Remember that salts of other metals, such as magnesium, calcium and potassium, are also present in sea water but only in very small quantities.

Can you answer these questions?

1. Why is chlorine gas produced at the anode during electrolysis?
2. What happens to the sodium when electrolysis of sea water is used to produce hydrogen and chlorine?
3. What are the main uses of sodium hydroxide in the home?

Did you know?

- Chlorine gas is so poisonous that it was used in World War 1 as a weapon.

Designer products

1 Smart materials

Have you ever wondered?

Why Gore-Tex™ is breathable?

Key facts

- New materials are being developed all the time. Sometimes scientists are trying to produce a material with specific properties, but occasionally a new material is discovered by chance.
- Lycra™: very flexible, body hugging and hard wearing: used in sports clothing.
- Thinsulate™: made from thin fibres, traps lots of air and heat; used in cold weather gear.
- Carbon fibres: as strong as steel but much lighter: used for sports equipment like fishing rods and tennis rackets.
- Shape memory alloys: atoms always return to their starting point, so even after the metal has been disturbed it can return to its original shape.
- Kevlar™: lightweight cloth that is fives times stronger than steel: used in body armour.
- Gore-Tex™: water vapour (gas) can get through but rain (liquid) cannot.
- Teflon: very slippery plastic that things cannot stick to: used in saucepans.

Definitions

- **Breathability** a property of clothing that lets the water vapour from sweat escape

❶ Examiner's tips

- As this is new technology, the materials have trade names and trade marks. Registering a trade mark helps the company that developed the material to make money. Remembering trade names is not as important as remembering the special properties of these materials and their uses.

Can you answer these questions?

1. If steel and Kevlar™ both do the same job, why do the police prefer Kevlar™?
2. Which materials have been accidental discoveries?

Look on the CD for more exam practice questions

Did you know?

- Sometimes, when an experiment goes wrong, a good scientist will still see an advantage. Post-it™ notes are an example of an experiment that did not work out as expected. The inventor was trying to make a strong permanent adhesive. Luckily, he realised the potential of a temporary adhesive and the removable Post-it™ note was produced.

② Nanoparticles

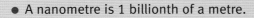

Have you ever wondered?

Are the new sunscreens that contain nanoparticles safe?

Key facts

- A nanometre is 1 billionth of a metre.
- This means it's too small to be seen, even with a microscope.
- A nanoparticle can vary in size from 10 to 100 nanometres.
- Nanoparticles are smaller than viruses, but bigger than individual atoms.
- Nanoparticles are used in sunscreens to reflect UV light.
- Silver nanoparticles are use to kill bacteria and fungi on toothbrushes, babies' dummies and medical equipment.
- Nanoparticles coat self-cleaning glass to stop organic molecules from sticking to the glass.
- Nanoparticles can make fabrics stain resistant.
- Iron nanoparticles can speed up the breakdown of oil and dry cleaning solvents.
- Nanocomposites use nanoparticles in combination with other materials.
- Nanotubes could be used as molecular sieves.

Definitions

- **Nanocomposite** a material that has nanoparticles used in combination with other materials

❶ Examiner's tips

- There are lots of nano-words, make sure you know what each one means.

Can you answer these questions?

1. Give three uses of nanoparticles.
2. What is a nanocomposite?.

Look on the CD for more exam practice questions

Did you know?

- Dr Who used medical nanobots to help repair human tissues in *"The Empty Child"*. These tiny little machines looked like fire flies but were so small they could repair human DNA and cells. Steven Moffat, who wrote this episode, read about nanobot technology and expanded the idea for the future!

③ Fermentation

Have you ever wondered?
How beer is made?

Key facts

- Yeast is a single-celled, living organism.
- It reproduces by budding.
- It uses sugars, usually from fruit or barley, to produce its energy anaerobically.
- This process it called **fermentation**.
- The waste products of fermentation are carbon dioxide and ethanol.
- Beer and wine are produced by fermentation.
- **Ethanol** is toxic, and eventually the yeast will be killed by their own waste products.
- **Alcoholic** drinks with a higher percentage of ethanol are distilled.
- Beer, wine and spirits all have different ethanol contents.
- Spirits, i.e. whisky, rum etc, contain about 35–40% ethanol,
- Wines contain about 12–14% ethanol.
- Beers contain about 3–7% ethanol.
- Alcohol is a depressant. This means it reduces the activity of the nervous system and slows down reaction times.
- Like many drugs, continued over-use can lead to dependency and addiction.
- Alcohol does not just affect individuals; fights, drunk driving and antisocial behaviour make drinking a problem for whole communities.

Definitions

- **Alcohol** a type of chemical made by fermenting sugars
- **Ethanol** the alcohol in alcoholic drinks
- **Fermentation** the process when yeast turns sugar into alcohol and carbon dioxide

❶ Examiner's tips

- No matter what alcoholic drink is being produced, the starting technique is the same. All that changes is the source of the sugar.
- The effects of alcohol can be separated into short term effects that happen to any one who drinks where the body can usually sort out and repair the damage, and long term effects that depend on the amount of alcohol you drink, how often you drink and the amount of time you have been drinking.

Can you answer these questions?

1. Why does fermentation need to be carried out in warm conditions?
2. Give two reasons why the yeast will eventually stop producing ethanol.

Look on the CD for more exam practice questions

Did you know?

- There are many kinds of alcohol but ethanol is the only one that is safe to drink (in moderation).
- Another alcohol, methanol, is very toxic and drinking even small amounts will cause permanent damage.

4 Intelligent packaging

Have you ever wondered?

How does intelligent packaging keep food fresh?

Key facts

- Microbes are constantly attacking our food.
- Packaging is used to keep the microbes from the food.
- Intelligent packaging is designed to go one stage further by controlling the environment surrounding the food.
- Oxygen is removed and replaced by unreactive gases like nitrogen.
- Lack of oxygen means microbes cannot grow or reproduce.
- Lack of oxygen prevents fruits turning brown.
- Lack of oxygen prevents fats going rancid.

❶ Examiner's tips

- Oxygen from the air is the main problem in food deterioration, but don't forget that moisture also plays a part.

Look on the CD for more exam practice questions

Can you answer these questions?

1. Why do manufacturers replace air with nitrogen, rather than just removing all the air from crisp packets?
2. Are there any disadvantages to intelligent packaging?

Did you know?

- Soon your food could talk to your oven; microchips in the packet would relay cooking information to control cooking times and temperatures!

⑤ Emulsions

Have you ever wondered?

How is the oil and water in mayonnaise kept from separating?

Key facts

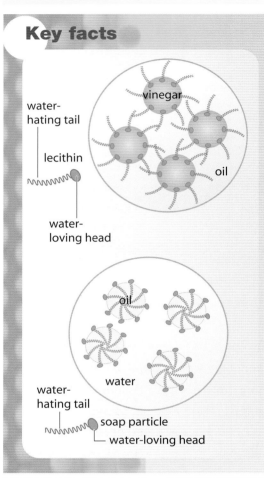

- An emulsion is a mixture of two immiscible substances, in which one is dispersed in the other.
- Emulsions tend to have a cloudy appearance.
- Emulsions are unstable and thus do not form spontaneously.
- Energy has to be put in to making them, for example, by shaking or stirring.
- If left, they will separate unless something is added to increase their stability.
- Emulsifiers are a group of substances that increase the stability of emulsions.
- Emulsifiers in foods are substances with molecules with two distinct ends. One that is attracted to water (hydrophilic) and one that repels water (hydrophobic)
- One end is attracted to the oil and the other to the water and this holds the two liquids together in the emulsion.
- Detergents work in the same way by making an emulsion of dirt and water and therefore attracting the dirt out of your clothes

Definitions

- **Emulsifier** a chemical that stops an emulsion from separating
- **Hydrophobic** the part of an emulsifier that repels water – it prefers oil
- **Hydrophilic** the part of an emulsifier that prefers water – it repels oil

❶ Examiner's tips

- Make sure you can name some common emulsions and some emulsifiers.

Can you answer these questions?

1. What does lecithin do?
2. How does making emulsions help to clean your clothes?

Look on the CD for more exam practice questions

Did you know?

- An emulsion is just one of a group of substances called colloids.

⑥ Designer products

Key facts

- To design new products, manufacturers must make a list of the essential and the desirable characteristics.
- The new product must have all the essential characteristics or people will not buy it.
- The desirable ones are those that make people choose one brand over another.
- Companies spend a lot of time and money on market research, gathering opinions and modifying their product before releasing it.
- Advertisers concentrate on what makes their product different from all the others.
- There are lots of rules and regulations to stop advertisers making exaggerated claims.

❶ Examiner's tips

- Think about the products you buy. When you choose products you are comparing brands. What makes the brand you chose better, for you, than the others? Next time you are in a supermarket look at a shelf and see how many brands there are for a single product. Look at how the manufacturers are trying to make their product stand out. Make a mental list of what you think is important and see how many brands support your ideas.

Look on the CD for more exam practice questions

Can you answer these questions?

1. Give three essential characteristics of a shampoo.
2. Give three desirable characteristics of a shampoo.

Did you know?

- In the year AD79 the ashes from burnt mouse heads were used to keep teeth clean and prevent toothache. Modern toothpastes do not list this as an ingredient!

Producing and measuring electricity

① Producing electric current

Have you ever wondered?

Which invention or discovery changed the world the most?

Key facts

- In a direct **current** (DC), electricity flows in one direction only.
- In an alternating current (AC), electricity flows first one way and then the other.
- A current can be generated in a **dynamo** by moving a magnet in a coil of wire.
- The direction of the current depends on the direction the magnet moves.
- In a generator the coil spins instead of the magnet. The coil is connected using brushes.

Definitions

- **Current** a flow of electricity around a circuit – a number of amps (A) or milliamps (mA)
- **Dynamo** a device that generates electricity when a magnet rotates inside a coil of wire

Look on the CD for more exam practice questions

❶ Examiner's tips

- The mains current at home is alternating current.
- The current from batteries is direct current.
- The voltage from a generator will be bigger if:
  - the magnet is stronger
  - the magnet moves faster
  - the coil has more turns of wire.

Can you answer these questions?

1. What is a dynamo?
2. What is the difference between DC and AC?
3. If the lights on your bicycle are powered by a dynamo, what happens to the lights when you:
   (a) speed up
   (b) stop?
4. Suggest three ways to reduce the voltage produced by a dynamo.

Did you know?

- The generator in a power station produces current just like a bicycle dynamo, but it can be as big as a house.

② Current, voltage and resistance

Have you ever wondered?

How a dimmer switch works?

Key facts

- The higher the **voltage**, the more energy the electricity has.
- Increasing the voltage increases the **current**.
- **Resistance** makes it difficult for a current to flow round a circuit.
- Increasing the resistance reduces the current.
- Voltage (V), current (I) and resistance (R) are related by the equation:

  $V = I \times R$

- Electrical current is a flow of negatively charged electrons.

Definitions

- **Voltage** the potential difference between two points – a number of volts (V) or millivolts (mV)
- **Current** a flow of electricity around a circuit – a number of amps (A) or milliamps (mA)
- **Resistance** how difficult it is for current to flow around a circuit – a number of ohms (Ω)

Look on the CD for more exam practice questions

❶Examiner's tips

- You should remember the equation that links voltage, current and resistance. This relationship is called Ohm's law.
- It is important to get the units right when you do any calculation. The units are:
  - voltage in volts
  - current in amps
  - resistance in ohms

Can you answer these questions?

1. What is the unit for current?
2. What is the unit for resistance?
3. What is the effect of increasing the voltage supplied to a circuit on:
   (a) the current in the circuit
   (b) the energy of the electricity that flows?
4 The voltage between the two ends of a piece of wire is 6 V and the current in it is 2 A. What is the resistance of the wire?

Did you know?

- The wires used to transmit electricity have to be very thick – up to a few centimetres across. This is because thicker wires have a lower resistance.

3 Resistance, lamps and computers

Have you ever wondered?

Whether a computer could do an experiment for you?

Key facts

- The resistance of a fixed **resistor** does not change unless the temperature changes.
- Warmer wires have a higher resistance.
- As a filament lamp warms up, its resistance changes a lot.
- The change can be shown by plotting a V/I graph like this.

V (V)	I (A)
0.00	0.00
1.50	0.40
3.00	0.78
4.50	1.00
6.00	1.15
7.50	1.28
9.00	1.42
12.00	1.62

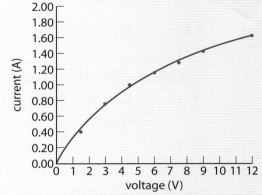

- At any point on the graph, the resistance can be found by using $R = V \div I$
- Sometimes computers are better at collecting data than people.
- Computers can work very quickly and don't get tired.

Definitions

- **Resistor** an electrical component that restricts the flow of electrical current – there are fixed-value resistors, and variable resistors
- **Ammeter** a meter for measuring electrical current

Look on the CD for more exam practice questions

❶ Examiner's tips

- The V/I graph for a lamp has a curved line because the lamp's resistance changes.
- The V/I graph for a fixed resistor has a straight line because the resistance is constant.
- An **ammeter** is always wired in series, it measures the current that flows through it

Can you answer these questions?

1. What is a resistor?
2. Sketch the V/I graph for a fixed resistor.
3. How does the resistance of a lamp change as it gets hotter.
4. Look at the graph above. Compare the resistance of the lamp at two different voltages.

Did you know?

- The Bell Rock lighthouse uses a 35 W lamp – that's only about half the power of a normal household light bulb. A big lens concentrates the light into a bright beam.

④ Changing resistance and controlling the current

Have you ever wondered?

How does my digital camera take great pictures automatically?

Key facts

- LDR means the same as **light-dependent resistor**.
- When it gets darker, the resistance of a LDR increases.
- When it gets lighter, the resistance of a LDR decreases.
- When it gets colder, the resistance of a **thermistor** increases.
- When it gets warmer, the resistance of a thermistor decreases.
- LDRs are used in light metering circuits, a change in resistance gives a change in current, e.g. in cameras.
- Thermistors are used in temperature monitoring circuits, a change in resistance give a change in current, e.g. in digital thermometers.

Thermistor   LDR

Definitions

- **Light-dependent resistor (LDR)** a resistor whose resistance changes with light intensity
- **Thermistor** a resistor whose resistance changes with temperature

Look on the CD for more exam practice questions

❶Examiner's tips

- Make sure you know the circuit symbols, so that you can recognise the components in a circuit diagram.
- When LDRs and thermistors are connected in series with the circuit, they can control the current.

Can you answer these questions?

1. What is the circuit symbol for a LDR?
2. What is a thermistor?
3. Name three devices that people use at home that contain a thermistor.
4. The display on a digital clock changes brightness automatically. It gets brighter in bright light and dimmer when the light fades. Suggest how this could be done.

Did you know?

- Most street lights have a circuit with an LDR that automatically switches the lamp on when it gets dark and off when it gets light again.

⑤ Cells and recharging

Have you ever wondered?

How can I make the batteries in my MP3 player last longer?

Key facts

- The current from batteries and solar cells is direct current (DC).
- The capacity of a **rechargeable battery** is given as a number of amp-hours.
- To work out how long a battery should last, you divide the capacity (in amp-hours) by the current in amps.
- Rechargeable batteries are more expensive than ordinary ones, but you can use them many times.
- Rechargeable batteries can help the environment because they can be used more than once before they need to be recycled.
- Some rechargeable batteries are very heavy.

Definitions

- **Rechargeable** a cell that can be used many times – a reverse current can restore its energy
- **Battery** several cells connected together

Look on the CD for more exam practice questions

❶ Examiner's tips

- Rechargeable batteries have disadvantages as well as advantages.
- If you are comparing rechargeable and ordinary batteries, try to make it a fair test and compare like with like.

Can you answer these questions?

1. What is a battery?
2. What is a rechargeable battery?
3. Why should batteries be recycled properly?
4. A battery is labelled 900 mAh (milliamp-hours). For how long should it run an MP3 player that needs a current of 45 mA?

Did you know?

- An electric milk float can deliver $1\frac{1}{2}$ tonnes of milk at a time – but it also needs to carry $1\frac{1}{2}$ tonnes of batteries.

6 Electricity and technology

Have you ever wondered?

How can a train possibly go at 500 kilometres per hour?

Key facts

- Having electricity at home lets us have energy whenever we need it.
- Having a telephone lets us communicate instantly over long distances.
- Electric **circuits** have become smaller. Now, even very complex equipment can be portable.
- When circuits are very small, the signals travel very short distances.
- This allows higher frequencies to be used, which can increase the processing power of a computer.
- When metals are very cold their resistance disappears – this is **superconductivity**.
- Superconductivity lets wires easily carry the huge currents needed by the electromagnets that can lift a train.

Definitions

- **Circuit** some electrical devices that have been connected together
- **Superconductivity** extremely low resistance that can happen at very low temperatures – electricity is conducted much better

❶ Examiner's tips

- A higher frequency can mean more processing power for a computer, but there is always a limit to the frequency that can be used – even with tiny circuits.
- Maglev trains can go very fast because floating magnetic suspension means that there is much less friction to overcome.

Can you answer these questions?

1. How did people communicate over long distances before they had telephones?
2. Name some forms of energy that we get from electrical equipment at home.
3. What are the main features of a Maglev train?

Look on the CD for more exam practice questions

Did you know?

- The earliest electronic computers were big enough to fill a room. Now a chip that is smaller than your fingernail has far more processing power.

You're in charge

1 Renewable energy versus fossil fuels

Have you ever wondered?

What would happen if all the electricity in the world went off and stayed off?

Key facts

- Oil, gas and coal are fossil fuels.
- Fossil fuels will run out. They are non-renewable resources.
- Energy from wind, water and the Sun is renewable.
- Renewable resources often cause little or no pollution.
- Renewable energy can be cheaper energy.
- Wind turbines are not a reliable source of energy – the strength of the wind varies.
- Solar panels are expensive to install, but they give cheap energy whilst the Sun is shining.
- Some people think that it is a bad idea to build wind turbines and electricity pylons because it spoils the countryside.

Definitions

- **Solar power** generating electricity using energy that comes to us from the Sun
- **Wind power** generating electricity using energy from wind turbines

Look on the CD for more exam practice questions

❶ Examiner's tips

- Renewable energy has disadvantages as well as advantages.
- Fossil fuels have advantages as well as disadvantages.
- All types of energy:
  - have some cost
  - can affect the environment
  - can affect the way we live.

Can you answer these questions?

1. Name three fossil fuels.
2. What is the difference between renewable and non-renewable resources?
3. Explain why a coal-fired power station causes pollution.
4. Why can we not rely on wind power for all our energy needs?
5. How can you find out if it is dangerous to live near high voltage electricity cables?

Did you know?

- People used to think that electricity could make you healthy; now some people are worried about living near electricity pylons. Ideas change over time.

② Motors

Have you ever wondered?

What kind of car will you be driving in 10 years time?

Key facts

- There is a **motor** effect on a coil of wire in a **magnetic field**.
- There is a force on the coil when current passes though it.
- The size of the force depends on:
  - the strength of the magnetic field
  - the size of the current
  - the number of turns on the coil.
- The direction of the force can be reversed by:
  - reversing the direction of the magnetic field
  - reversing the direction of the current.
- The coil in the motor is connected using brushes and a commutator, this helps the motor to keep spinning. This diagram shows a simple electric motor.

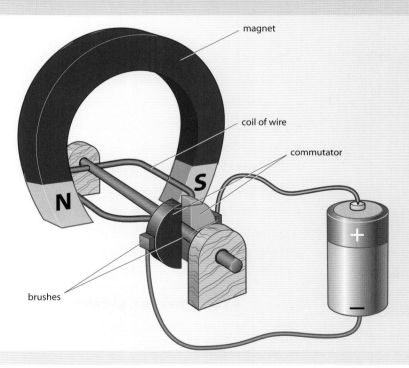

Definitions

- **Magnetic field** the space around a magnetic object, it affects other magnetic objects
- **Motor** something that turns one form of energy into movement

❶ Examiner's tips

- Swapping over the battery connections reverses the current.
- Swapping over the magnets reverses magnetic field.
- Reversing both the current and the magnetic field together will not affect the way the motor spins.

Look on the CD for more exam practice questions

Can you answer these questions?

1. What household items use an electric motor?
2. How can you change the direction that an electric motor turns?
3. Which factors affect the size of the force on the coil of an electric motor?
4. A radio controlled toy car uses a d.c. motor. Why?

Did you know?

- The smallest electric motor is about 500 nm across. That's 300 times smaller than the diameter of a human hair.

③ Wires, fuses and safety

Have you ever wondered?

Will a 240V electric shock kill you?

Key facts

- The earth wire connects the metal case of an appliance to earth.
- If there is a fault and a live wire touches the case, a large current flows.
- This large current is enough to blow the **fuse** and cut the circuit.
- This stops a fault overheating the wires and starting a fire.
- You feel an electric shock if a current more than 1 mA passes through you.
- If some of the current from the live wire goes through you, it does not go back through the neutral wire.
- A **residual current circuit breaker (RCCB)** detects the small difference in current between live and neutral wires.
- If this difference is more than about 30 mA the RCCB breaks the circuit.

Definitions

- **Fuse** a device in a mains plug that protects equipment and cables from excess current and the resulting fire hazard
- **Residual current circuit breaker (RCCB)** a device that shuts down an electrical circuit to protect people from electric shock

❶ Examiner's tips

- Fuses protect the wires from overheating. Hot wires might cause a fire.
- If you touch a live wire, a fuse might not save you. A current of only 100 mA can be fatal – most fuses blow at much higher currents.
- With a RCCB, you might get a shock, but the current stops before you are injured.

Look on the CD for more exam practice questions

Can you answer these questions?

1. What is a fuse?
2. A fuse is labelled 5 A. What does the label mean?
3. What is the best way to avoid receiving an electric shock when
    (a) touching the metal case of a washing machine
    (b) using a mower that might cut the mains cable.
4. Brian was using an electric drill when he accidentally drilled into a mains cable in the wall. Brian did not get a shock, but the drill stopped working and the lights went out. What does this tell you about the electrical equipment?

Did you know?

- RCCBs usually operate within 25 milliseconds. This is quick enough to prevent a person's heart being damaged or stopped by an electric shock.

4 Power

Have you ever wondered?

How many devices can you safely plug into one wall socket?

Key facts

- The more powerful something is, the faster it can transfer energy.
- Electrical energy can be converted to different types of energy; the rate of transfer is still called the **power**, whatever the type of energy.
- A 60 W light bulb transfers energy more quickly than a 25 W light bulb.
- A 1 kW toaster transfers energy less quickly than a 2 kW toaster.
- The equation for calculating electrical power is:

  power = current x voltage
- This can be rearranged to find current from power and voltage:

  current = power ÷ voltage

Definitions

- **Power** how quickly something converts energy from one form to another – a number of watts (W) or kilowatts (kW)

❶ Examiner's tips

- Electrical power is a number of watts or kilowatts.
- Appliances designed to produce heat usually have higher power ratings – e.g. an electric kettle might be rated 2 or 3 kW.

Look on the CD for more exam practice questions

Can you answer these questions?

1. What is power?
2. What are the units for power?
3. The current in a toaster is 5 A and the voltage is 230 V. What is the power of the toaster?
4. A kettle is labelled 2.3 kW, 230 V. What fuse should it have?

Did you know?

- We use 13 A plugs because very few household appliances exceed 3 kW in power. (3000 ÷ 230 = 13)

⑤ Paying for electricity

Have you ever wondered?

How can you save electricity and reduce your costs?

Key facts

- Your **electricity** bill asks you to pay for the amount of **energy** you used.
- This amount energy is measured as a number of kWh (kilowatt-hours).
- The number of kWh used depends on the power (in kW) and the time (in hours).
- The cost of 1 kWh can vary (depending on fuel costs and whether the energy was used at an off-peak time or not).
- The formula for calculating the cost of electrical energy is:

  cost = power × time × cost of 1 kWh

- Energy-efficient measures usually have costs and savings.
- If you divide the cost by the yearly saving, you get the pay-back time.

Definitions

- **Electricity** the energy involved when charged particles flow
- **Energy** whenever something happens, energy changes from one form to another – a number of joules

❶ Examiner's tips

- Expensive energy-efficient measures, like double glazing, can have long pay-back times.
- If you are given the cost of 1 kWh as a number of pence, then the cost you calculate will be a number of pence too.

Can you answer these questions?

1. What is the unit for buying electrical energy?
2. Why is it cheaper to boil a kettle of water when it is half full than when it is completely full?
3. Arvind uses a 3 kW heater for 6 hours. 1 kWh costs 12p. What was the cost of using the heater?
4. Davina spends £6000 on double glazing. It saves her £300 a year in fuel costs. What is the pay-back time?

Look on the CD for more exam practice questions

Did you know?

- In some areas of the UK, the local government will give grants to help pay for the cost of improving home insulation.

6 Solar cells and efficiency

Have you ever wondered?

Could your bedroom be powered by renewable energy?

Key facts

- **Efficiency** is a measure of how good something is at converting one sort of energy to another.
- The formula for finding efficiency is: $\frac{\text{useful output}}{\text{total input}} \times 100\%$
- Unless something is designed to produce heat, it usually wastes energy as heat or sound – the more energy wasted the lower the efficiency.
- **Solar cells** are expensive to buy, but they can provide very cheap energy.
- A small solar cell can easily power something like a calculator, but for something like a washing machine, a very large area of solar panels would be needed.

Definitions

- **Efficiency** a measure of how good something is at turning energy from one form into another useful form
- **Solar cell** a cell that produces voltage by converting light into electrical energy

❶ Examiner's tips

- Efficiency is a ratio of two amounts of energy – it has no unit.
- Efficiency is never more than 100% – the useful output is always less than the total input.
- If you are comparing different solar cells, try to make it a fair test and compare like with like.

Can you answer these questions?

1. What is efficiency and why does it not have a unit?
2. Why is a low-energy light bulb more efficient than an ordinary one?
3. A guitar amplifier turns 100 J of electrical energy into 40 J of sound. What is its efficiency?
4. Explain why solar cells are rarely used at home, but are often used to power satellites in space

Look on the CD for more exam practice questions

Did you know?

- Even the best solar cells are unlikely to be more than 20% efficient – but this does not matter very much, because energy from the Sun is free.

Now you see it, now you don't

1 Wave basics

Have you ever wondered?

Why are there so many different types of waves?

Key facts

- Waves transfer energy from place to place and can differ in **amplitude**, **wavelength**, frequency and speed.
- Frequency (f) and wavelength (λ) are related to the speed (v) at which energy moves $v = f \times \lambda$
- **Electromagnetic spectrum** waves (unlike sound) can travel through a vacuum and all have the same speed there.
- Waves can be **longitudinal** or **transverse**. Electromagnetic waves are transverse.
- Digital signals have advantages over analogue signals.

Definitions

- **Amplitude** the maximum distance that particles in a wave move from their normal positions
- **Wavelength** the distance a wave moves during the time for one full vibration
- **Longitudinal** a wave where the vibration is parallel to its direction of travel
- **Transverse** a wave where the vibration is at right angles to its direction of travel
- **Electromagnetic spectrum** a group of transverse waves that all travel at the same speed in a vacuum

❶ Examiner's tips

- For a given speed, the higher the frequency of a wave the shorter the wavelength
- Up until the start of the 20th century, the only waves that people deliberately used were light and sound. Now things are very different.
- There are benefits and drawbacks to our reliance on these other waves.

Can you answer these questions?

1. Name a wave which is **not** transverse.
2. What happens to the speed of a wave if its wavelength increases?
3. Ultraviolet light has a higher frequency than microwaves. What can you say about their speeds?
4. State, in sentences, the information given by this diagram:

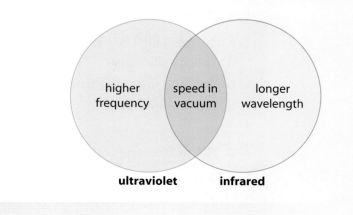

Look on the CD for more exam practice questions

Did you know?

- Low frequency radio waves and high frequency gamma rays travel through space at exactly the same speed.

② Reflections

Have you ever wondered?

How do you see an unborn baby?

Key facts

- Reflection of light from materials of different density helps us see most objects.
- The time for travel, reflection and return of a light ray indicates the distance away from the object. For example, the depth of fish in water or the distance between an airport and a plane.
- Distance = speed x time
- **Ultrasound** is used to scan unborn babies.
- Optical fibres use total internal reflection.

Definitions

- **Scanning** moving a detector across something to build up an image
- **Ultrasound** sound waves with a frequency too high to hear

❶ Examiner's tips

- When working out distances by reflection, the time to use is *half* the time measured because the wave has to travel there *and* back.
- Various parts of the body have different densities and so cause reflections with ultrasound.
- Ultrasound scans are good because energy is reflected and not absorbed by the unborn baby.

Can you answer these questions?

1. Show, by a diagram, how light passes along an optical fibre.
2. Name the process used in reflecting light in optical fibres.
3. Explain how barcodes are scanned in shops.
4. A student shouts at a cliff 100 m away. He hears an echo 0.6 s later. What is the speed of sound?

Look on the CD for more exam practice questions

Did you know?

- It takes light just over $2\frac{1}{2}$ seconds to be reflected off the Moon and back to the Earth.

③ Scanning by absorption and emission

Have you ever wondered?

How can microwaves be used to forecast the weather?

Key facts

- Scanning by infra-red **emission** can be used to measure temperatures.
- Astronomers scan the sky by the light emitted by stars.
- A gamma ray source carried by oil can be used to detect oil leaks.
- The flow of liquids in the body, and **absorption** of some chemicals, can be traced by radioactive emission.
- Some wavelengths of microwaves are absorbed by water, which heats the water.
- X-ray absorption can detect broken bones.
- Ultraviolet light absorption helps in the detection of forgery using **fluorescence**.

Definitions

- **Emission** producing and sending out waves
- **Absorption** when a material absorbs some energy from a wave travelling in it
- **Fluorescence** when a substance absorbs ultraviolet light and emits visible light

❶ Examiner's tips

- Absorbing X-rays is dangerous.
- Scanning by emission uses radiation produced by the object itself.
- Scanning by absorption involves some objects removing energy from the beam.
- Microwave absorption is used for detecting rain clouds.

Can you answer these questions?

1. Name one substance which absorbs X-rays.
2. What is the difference between emission and absorption?
3. Explain what happens to a wire as it is gradually heated until it glows white hot.
4. Why is an ultrasound scan safer to use than an X-ray?

Did you know?

- Scientists can tell the temperature of stars by the colour of the light they emit.

Look on the CD for more exam practice questions

④ Digital signals and mobile phones

Have you ever wondered?

Why is the picture better on a digital TV?

Key facts

- Mobile phones have advantages over land lines.
- Microwaves are used in mobile phones.
- Some microwave wavelengths are used in cooking as they cause internal heating.
- There is concern that radiation from radio masts may cause health problems, but there is no proof at present.
- **Digital** signals have only two values which are often shown as 1/0 or on/off.
- Music can be synthesised by adding different digital waves.

Definitions

- **Digital** a signal that has only two levels, e.g. on or off
- **Analogue** a signal that varies continuously, e.g. a wave

❶Examiner's tips

- Newspapers, TV and radio present problem issues, such as the safety of mobile phones, in different ways.
- The presentation may be biased by industrial/economic interests, politics or other prejudices.
- Some just appeal to the emotions while others present evidence and then argue conclusions from this.
- There are advantages and disadvantages to mobile phones.
- There are advantages in using digital signals rather than analogue.
- New music technologies have brought about many changes, for better or worse, in how, when and where we listen to music.
- Many modern entertainment, medical and industrial technologies rely on a variety of waves. Consider the advantages and disadvantages of these to both yourself and your locality.

Look on the CD for more exam practice questions

Can you answer these questions?

1. Use words from the box to complete the sentences below.

   | longitudiual analogue transverse digital |

   (a) Vinyl records store information in .... format.

   (b) Music is synthesised by adding waves which are ....

2. Which parts of the electromagnetic spectrum have a wavelength shorter than visible and a frequency less than gamma rays?

Did you know?

- X-rays and gamma rays can have the same wavelength. The only difference between them is the way they are produced.

5 How dangerous are these waves?

Have you ever wondered?

Why does your skin burn more quickly in the midday sun?

Key facts

- There is no proof at present about the dangers of radio waves.
- One particular frequency of microwaves makes water molecules vibrate and so heats internal body organs.
- Infrared is used to make toast and can burn the skin.
- Visible radiation (e.g. sunlight) can damage eyes.
- Ultraviolet radiation can damage eyes and cause sunburn and skin cancer.
- X-rays can kill cells deep in the body and cause mutations.
- γ-rays (gamma rays) can kill cells deep in the body and cause mutations.

Definitions

Mutation when the DNA of cells is altered

❶ Examiner's tips

- Electromagnetic waves are useful. For example, ultraviolet waves from either the Sun or a sun-bed can help in the formation of vitamin D as well as giving you a suntan, but be aware of the dangers.
- The greater the exposure to any wave the greater the danger. Potentially, however, the danger of radiation increases with increasing frequency.

Look on the CD for more exam practice questions

Can you answer these questions?

1. Draw a line from each part of the body to the wave most likely to damage it.

Part of body	Type of wave
Eyes	microwaves
Internal organs	ultraviolet
Cells deep in body	
Skin	gamma

2. Explain why X-rays are potentially more dangerous than infrared waves.

Did you know?

- Amount does matter! A little is often great, but too much can kill.

6 Earthquake waves

Have you ever wondered?

Why do scientists believe there could be an even more catastrophic tsunami than the one in the Indian Ocean on 26 December 2004?

Key facts

- Earthquakes are caused when large chunks of the Earth's surface slide against each other producing **seismic waves**.
- Primary (P–) waves travel faster than Secondary (S–) waves.
- P–waves are longitudinal and S-waves are transverse.
- Transverse waves cannot pass through liquids.
- Earthquakes can trigger **tsunami**.
- Seismic waves help us see inside the Earth.
- Scientists cannot yet predict earthquakes but can give the chance of them occurring.

Definitions

- **Seismic wave** a wave produced by an earthquake
- **Tsunami** a wave produced by an undersea earthquake or volcanic eruption

❶Examiner's tips

- The Earth is made up of several layers like an onion. Scientist say that there is a liquid core at the centre of the Earth because transverse waves cannot be detected at parts of the Earth's surface.

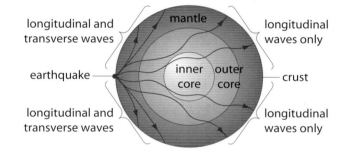

- Seismic waves are bent (refracted) due to change of density.
- Scientists cannot yet predict when earthquakes will happen, owing to their complexity.

Look on the CD for more exam practice questions

Can you answer these questions?

1. (a) Name the two types of seismic waves abbreviated as P– and S–.
   (b) How do they differ?
2. What is the difference between a seismic wave and a tsunami?
3. Explain why only one type of seismic wave (P or S) would be detected at the south pole if an earthquake happened at the north pole.

Did you know?

- In about AD132, Chinese scientists detected earthquakes using a circular device around which were placed model dragons, each holding a ball in its mouth. Shock waves moved a pendulum which opened the jaws of the dragon facing the direction of the earthquake. A ball fell from the dragon's teeth into the mouth of a model toad below to record the event.

Space and its mysteries

1 A weighty problem

Have you ever wondered?
What did Newton learn from an apple?

Key facts
- Everything in the Universe attracts everything else.
- Planets, comets, asteroids etc. orbit the Sun due to gravity.
- All objects are held on the Earth by the force of gravity.
- An object's **mass** is the same everywhere in the Universe.
- The **weight** of an object differs from place to place.
- Weight = mass x acceleration of free-fall ($w = mg$)

Definitions
- **Mass** how much of something there is – a number of kg
- **Weight** force exerted on an object by gravity – a number of N

Look on the CD for more exam practice questions

❶ Examiner's tips
- Mass is measured in kilograms, kg.
- Weight is a force measured in newtons, N.
- The unit of g (gravitational field strength) is newton/kilogram or N/kg
- Gravity does not suddenly stop at a particular height but it does reduce gradually as you go upwards.

Can you answer these questions?
1. What quantity is measured in kilograms?
2. A spaceship weighs 2 million newtons as it takes off. What is the minimum (smallest) force that the rocket engine needs to use?
3. The graph shows how the weight of an object changes with height above the Earth.

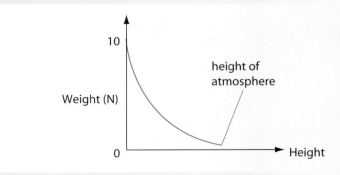

Sketch how you think the graph will continue.

Did you know?

- Your hands are as heavy as your feet if you measure them like this!

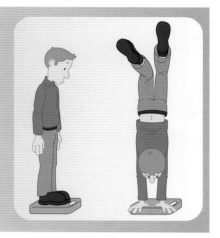

2 Where do we live and who are our neighbours?

Have you ever wondered?

The risk of dying from an asteroid impact is the same as being in an air crash. How can this be?

Key facts

- The Solar System consists of objects moving round the Sun.
- The Sun is one of millions of stars in our galaxy – the Milky Way.
- Planets are in roughly circular orbits around a star and moons orbit some planets.
- Asteroids are lumps of rock in orbit between Mars and Jupiter.
- Comets have long thin elliptical orbits around the Sun.
- The chance of collision between a comet and a planet depends, for example, on their size.

Definitions

- **Asteroid** a solid rock in space – about 10 to 500 km across
- **Comet** a ball of ice (and dust) in a very elliptical orbit around the Sun

❶ Examiner's tips

- Moons orbit planets and planets orbit stars along with asteroids and comets.
- Large groups of stars make up galaxies and large groups of galaxies make up the Universe.

Can you answer these questions?

1. Which planet has an orbit between Earth and Jupiter?

2. The pie chart below corresponds to the total mass of all the planets in the Solar System.

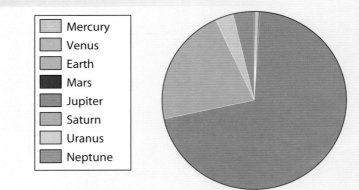

Mercury
Venus
Earth
Mars
Jupiter
Saturn
Uranus
Neptune

(a) What does the pie chart show about the mass of Jupiter?

(b) Why is it difficult to compare the masses of Mercury and Venus using this pie chart?

3. Describe the differences between the orbits of comets and planets.

4. Imagine that a scientist friend tells you that she has just discovered a new planet. What would you do?

Did you know?

- Comets are one of the few astronomical objects which change shape considerably during an orbit.

❸ I'm an Earthman – get me out of here

Have you ever wondered?

Is it worth £25 billion to put astronauts on Mars, when we could just send robots?

Key facts

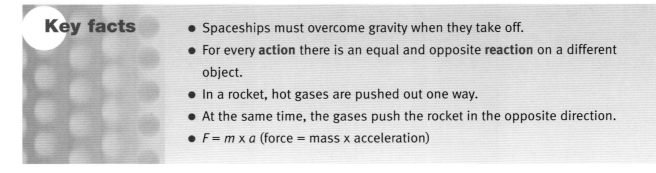

- Spaceships must overcome gravity when they take off.
- For every **action** there is an equal and opposite **reaction** on a different object.
- In a rocket, hot gases are pushed out one way.
- At the same time, the gases push the rocket in the opposite direction.
- $F = m \times a$ (force = mass x acceleration)

Definitions

- **Action** forces come in pairs – one of the forces is called the action
- **Reaction** forces come in pairs – the opposing force is called the reaction

Look on the CD for more exam practice questions

❶ Examiner's tips

- Action and reaction are equal in size but act on different objects.
- Action and reaction are in opposite directions.
- The same action/reaction effect is used for changing direction on a space walk.
- Rockets do *not* need anything to push against.

Can you answer these questions?

1. In what units are forces measured?
2. Imagine you are on a space walk. Explain how you could start to move backwards.
3. The same force acts on two objects. Explain why their accelerations are different.
4. During part of its flight, the shuttle has an acceleration of 3 m/s^2 when its mass is 2.3 million kg. What is the force in newtons at this time?

Did you know?

- To walk forwards, you push the Earth backwards. Don't try it when you get out of a small boat!

④ What's it like out there?

Have you ever wondered?

What is weightlessness like?

Key facts

- Each planet has a different gravitational field strength.
- Planets at greater distances from the Sun receive less radiation.
- Some planets have atmospheres.
- Some planets have moons.
- Artificial gravity can help us exercise in space.
- Special conditions are needed for human flights.

Definitions

- **Weightlessness** when there is no gravitational field

❶ Examiner's tips

- Whether a planet has the correct conditions for life depends partly on the size and temperature of the Star and the planet's distance from it.
- Space travel is expensive but discoveries may compensate for this.
- Space research so far has produced many technological advances and knowledge which have had social and economic benefits.

Look on the CD for more exam practice questions

Can you answer these questions?

1. On the Moon, would you expect the high jump record to be more or less than on Earth? How sure are you?

2. An object has a weight of 200 N on Earth where the gravitational field strength is 10 N/kg. It is sent to Jupiter where the gravitational field strength is 23 N/kg.

   What is

   (a) its mass on Earth?

   (b) its mass on Jupiter?

   (c) its weight on Jupiter?

3. John saw a star and two planets lined up like this.

   (a) Which planet would probably have the lower temperature?

   (b) If Q takes 2 years for each orbit and P takes 3 years, how long will it be before they look like this again?

   (c) How long will it be before they are lined up again on opposite sides of the star?

4. Name three conditions in space which are different to those on Earth.

Did you know?

- Scientists are developing materials for space suits which will repair themselves automatically.

⑤ Is anybody else there?

Have you ever wondered?

The Universe is full of planets where intelligent life could start, so where is everybody?

Key facts

- No human has ever been to another planet yet but unmanned probes using data loggers have.
- The **Universe** has existed for about 15 billion years.
- **SETI** (the Search for Extraterrestrial Intelligence) looks for patterns in signals from space using telescopes which receive radio waves, visible light, infra-red, etc.
- Problems with detecting life elsewhere include the immense distances between stars.

Definitions

- **Universe** all of space – it includes every galaxy
- **SETI (the search for extraterrestrial intelligence)** an organisation that seeks life forms from other planets by searching for signals

Look on the CD for more exam practice questions

❶Examiner's tips

- Unmanned probes don't need special conditions.
- Astronomers have discovered other planets but so far have not found any signs of life.
- Humans may be able to travel to other planets in the Solar System and they can then decide how and where to look for life.
- Life is not necessarily intelligent enough to send signals.

Can you answer these questions?

1. What does SETI stand for?
2. Give **one** advantage of sending a human to a planet rather than an unmanned probe.
3. Give **five** advantages of sending an unmanned probe to a planet rather than a human.
4. Earth has been emitting radio signals for about 100 years. Imagine that we receive a reply tomorrow. What is the maximum distance (in light years) that the alien could be from us? (1 light year is the distance light travels in 1 year.)
5. Imagine you met an alien while walking alone in the countryside. What would you do?

Did you know?

- SETI are looking for patterns with a scientific meaning. An artist claims to have detected signals showing an alien work of art.

⑥ Where do we go from here?

Have you ever wondered?

Do physicists really have no idea what most of the Universe is made from?

Key facts

- Stars do not live forever.
- The Universe is thought to have started with a **Big Bang** but there are other theories.
- **Red shift** and the presence of background radiation are evidence for the Big Bang theory.
- Ultimately, the fate of the Universe will depend on its mass.
- Scientists think that the Universe contains dark matter which is massive.
- Stars like our Sun will become red giants, then white dwarfs and finally black dwarfs.

- Stars of greater mass than our Sun will become neutron stars or black holes.
- A black hole has such a strong gravitational field that even light cannot leave it.

Definitions

- **Big Bang** theory suggests that this Universe came into existence in an explosion at the beginning of time
- **Red shift** the observed change in waves emitted by objects that are moving away – a decrease in frequency and an increase wavelength

Look on the CD for more exam practice questions

❶ Examiner's tips

- The graph illustrates three possibilities that scientists think could happen if the Universe has various masses.

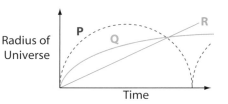

- At present, scientists cannot account for sufficient mass but think there may be some dark matter which will be enough to make up the difference.

Can you answer these questions?

1. Describe the changes in the size of the Universe shown in graph **P** above.
2. Label each diagram to show the stages in the evolution of a star like our Sun.

3. How does the Red shift give evidence for the expansion of the Universe?

Did you know?

- The amount of red shift increases as the speed increases.

Equations, units and symbols

These equations will be given to you when you need them.

P1a.9 Producing and measuring electricity

$V = I \times R$ (voltage, current and resistance)

P1a.10 You're in charge

power = current $\times$ voltage

efficiency = $\dfrac{\text{useful output}}{\text{total input}} \times 100\%$

(You need to memorise this equation)

cost = power $\times$ time $\times$ cost of 1 kWh

P1b.11 Now you see it, now you don't

speed = frequency $\times$ wavelength

speed = $\dfrac{\text{distance}}{\text{time}}$

P1b.12 Space and its mysteries

$w = m \times g$

weight = mass x acceleration of free-fall

$F = m \times a$

force = mass x acceleration

You should know and be able to use these units.

volt (V)
amp (A)
ohm (Ω)
watt (W)
kilowatt (kW)
joule (J)
kilowatt hour (kW h)
metre/second (m/s or m s^{-1})

newton/kilogram (N/kg or N kg^{-1})
newton (N)
kilogram (kg)
second (s)
metre (m)
milliamp (mA)
millivolt (mV)
metre/second/second (m/s^2 or m s^{-2})

You should know and be able to use these symbols.

ammeter	—(A)—	milliammeter	—(mA)—
voltmeter	—(V)—	millivoltmeter	—(mV)—
LDR		thermistor	
battery		cell	
power supply	(DC) or (AC)	fixed resistor	
variable resistor		switch	
fuse		lamp	

Glossary

B1a.1 Environment

adaptation how organisms change to suit to their environment better

biomass the mass of an organism (or a population) after all the water is removed

breeding mating organisms to increase the population size

characteristic a feature, like skin colour, which is either inherited or modified by the environment

classification a way to group things systematically

competition where two organisms both try to take the same thing, e.g. food, space, light

ecosystem a group of plants and animals that interact

environment the place where an organism lives, e.g. air, water, soil, other living things

evolution a change in characteristics, gradually over many generations

extinct a species that died out because it could not adapt to a new situation

food chain a chain showing how energy moves between plants and animals

fossil a rock imprinted with the body of an organism from millions of years ago

genetic engineering changing genes in an organism by altering its DNA or by transferring genes from another organism

genetically modified an organism with characteristics altered by genetic engineering

interdependence a change in one organism causes a change in another

inter-species between two *different* species

intra-species between members of the *same* species

natural selection when environmental factors like disease or predation change the characteristics of a species

organic food is produced without artificial fertilisers, pesticides or herbicides

organic farming uses natural ways of controlling pests and keeping soil fertile

organism a living thing

population the total number of a single species living in an area

predator an animal that eats other animals, e.g. cheetah

prey an animal that is killed for food by a predator, e.g. zebra

quantitative describing something by using numbers

reproduction propagation of the next generation

species a population of organisms which breed together to produce fertile young

B1a.2 Genes

allele alternative forms of the same gene – e.g. the flower colour gene in peas has alleles for white and yellow

antibody a substance white blood cells produce in response to infection that locates disease-causing organisms so they can be destroyed

asexual reproduction reproduction where only one parent passes on genes

cancer rapid uncontrolled growth of cells to form tumours which may spread

cell the basic unit of life – a nucleus and cytoplasm surrounded by a membrane

characteristic a feature of an organism e.g. flower colour or blood group

chromosome a thread-like string of genes in the nucleus

clone genetically identical plants or animals produced asexually from one parent

cystic fibrosis a genetic disorder that makes a person produce abnormal, sticky mucus in their lungs

DNA deoxyribonucleic acid – the chemical code that governs cell development

dominant an allele that overrides other alleles of a gene, hiding their effects

environment the surrounding conditions in which an organism develops

fertilisation when two sex cell nuclei join together, e.g. egg and sperm

forensics using scientific knowledge to help detect crime

gene a piece of DNA with the instructions for a particular characteristic, e.g. eye colour

generation the descendents of a pair of individuals – the first generation is their children, the second generation is their grandchildren

genetics studying the way characteristics are inherited

Human Genome Project a project that worked out the sequence of genes in human DNA

inheritance the passing of genes from parents to offspring

nucleus the part of a cell that contains the chromosomes

recessive an allele that is overridden by a dominant one – its effects only show when it is inherited from both parents

sexual reproduction when half the genes are inherited from each of two parents

transgenic an organism containing genes taken from another species

transplant an organ that was donated by one organism and inserted into the body of another - e.g. kidney transplant

variation the differences shown in a group of organisms, e.g. fur colour, flower colour

B1b.3 Electrical and chemical signals

bacteria a type of microorganism – some cause disease, but most are useful

brain an organ that coordinates the actions of the body

central nervous system (CNS) the brain and spinal cord

contraception a way to prevent pregnancy

diabetes a disease caused by uncontrolled sugar levels in blood

electrical impulse a signal carried by the nerves

epilepsy too much electrical activity in the brain – causes a seizure

genetically modified an organism with characteristics altered by genetic engineering

grand mal epilepsy that causes a person to lose consciousness and make jerky movements

hormone a chemical message that coordinates the body – produced by a gland

infertility when a couple cannot have children, e.g. blocked oviduct, insufficient sperm

insulin a hormone that controls the level of sugar in the blood – made in the pancreas

in-vitro fertilisation (IVF) the fertilisation of a human egg outside the body

iris reflex a reflex action that controls how much light enters the eye

menstrual cycle the cycle that prepares the uterus to receive a fertilised egg after ovulation

muscle tissue made from cells which can contract - allows movement to take place

oestrogen a hormone that thickens the uterus lining and stops eggs from developing

pancreas an organ in your abdomen that produces insulin

Parkinson's disease a disease that makes the brain unable to coordinate muscles properly

pregnancy the development of an embryo – from fertilisation until birth

progesterone a hormone that thickens the uterus lining after ovulation

reaction time the time that your body takes to react to a stimulus

receptor a cell that detects a stimulus, e.g. light, sound, heat

reflex an automatic response to a stimulus – cannot be controlled

sense organ an organ that has receptors to detect stimuli

stimulus something you react to, e.g. sound, heat, light (plural: stimuli)

stroke a blood clot or bleeding in the brain that kills brain cells

target organ the organ on which a particular hormone works

testosterone the male sex hormone

tumour cells growing to form an abnormal tissue

voluntary response a non-automatic response to a stimulus – can be controlled

B1b.4 Use, misuse and abuse

addiction a dependency on a drug

alcohol a sedative drug – produced by yeast

antibody a protein that destroys a pathogen – produced by white blood cells

antigen a protein marker on the surface of a cell – identifies it as belonging to the body or a foreign body

bacteria a type of microorganism – some cause disease, but most are useful

barbiturate a sedative that slows down the nervous system

barrier it keeps things out, e.g. the skin is a barrier against microorganisms

caffeine a stimulant drug – increases alertness

cannabis an illegal drug – may relieve pain

cilia tiny hair-like structures that clear mucus from the lungs and windpipe

circulatory system your heart, blood vessels and blood

disease caused by microorganisms – it makes you ill

drug a chemical designed to change the way your body works

foreign body something which is not part of the body

gaseous exchange the way your lungs take in oxygen and expel carbon dioxide

immune system the body's system for destroying invading microorganisms

infection when microorganisms invade the body

inflammation a response to infection – causes swelling and redness

lysozyme an enzyme found in tears that destroys bacteria – acts as a chemical barrier

microbe tiny organism – same as microorganism

microorganism tiny organism – like bacteria, viruses and fungi

neurone a single cell that carries impulses

opiate a drug from the opium family, e.g. opium, heroin

organism a living thing

overdose taking more than the recommended amount of a drug (may lead to death)

pain-relief a method that blocks impulses that tell the body about pain

paracetamol a general-purpose painkiller

pathogen a microorganism that causes disease

reaction time the time that your body takes to react to a stimulus

sedative a type of drug – can slow the nervous system down

solvent a chemical used to dissolve other substances, e.g. aerosols, thinners (may be abused by sniffing)

stimulant a drug that increases alertness and heart rate

tobacco a drug made from the tobacco plant and smoked – contains nicotine

transmission passing from one thing to another

tuberculosis a disease that causes damage to lungs – caused by a bacterium (may lead to death)

vector-borne disease carried by an insect, e.g. malaria is carried by a mosquito

viral infection an infection caused by a virus – antibiotics cannot cure it

white blood cell a type of cell in blood – part of the immune system

C1a.5 Patterns in properties

alkali metal an element from group 1 of the periodic table, e.g. lithium, sodium and potassium

analytical analytical tests find out what is present in a sample

atomic number the number of protons in the nucleus of an atom

atoms the smallest particles of an element that can exist

chemical symbol a code that represents an atom of an element – mostly one or two letters

compound atoms of two or more different elements chemically joined to form a substance

diatomic molecule two atoms joined by chemical bonds to make a molecule

electron a negatively charged particle that surrounds the nucleus of an atom

elements substances that are made of only one type of atom

endothermic a reaction that takes in heat from the surroundings

exothermic a reaction that gives out heat to the surroundings

flame test an analytical test to find out if a compound contains a metal – different metals produce different colours in a Bunsen flame

formula a code for a compound – uses symbols to show the ratio of atoms of the different elements it contains

group a vertical column of elements in the periodic table

halogen an element from group 7 of the periodic table, e.g. chlorine, bromine, iodine

inert unreactive

molecule atoms joined by chemical bonds to make a particle – can be an element or a compound

neutron an electrically neutral particle in the nucleus of most atoms

noble gas an element from group 0 of the periodic table, e.g. helium, neon, argon

period a horizontal row of elements in the periodic table

precipitation a reaction where two solutions are mixed together to form a solid

proton a positively-charged particle in the nucleus of all atoms
solution the mixture formed when a substance dissolves in a liquid – usually water
symbol *see* chemical symbol
transition metal an element from a group between groups 2 and 3 of the periodic table, e.g. copper, iron and gold

C1a.6 Making changes

carbohydrates compounds that contain carbon, hydrogen and oxygen only
caustic soda the common name for sodium hydroxide
citric acid an acid found in oranges and lemons
combustion a chemical reaction between fuel and oxygen – the burning process gives out heat
decomposition a reaction where a substance breaks down to form two or more new substances
dehydration a reaction where water is removed
dilute less concentrated – mixed with water
hydration a reaction where water is added
insoluble salt a salt that does not dissolve in water
neutralisation a reaction where an acid and an alkali (or other base) form a neutral solution
oxidation a reaction where oxygen is added to an element or compound
precipitate the insoluble solid formed in a precipitation reaction
salt one of the substances formed in a neutralisation reaction – a compound where a metal has replaced the hydrogen in an acid
soluble salt a salt that dissolves in water
thermal decomposition a reaction where thermal energy breaks a substance into new substances

C1b.7 There's one Earth

acid rain rain containing dissolved acidic pollutants, e.g. sulphur dioxide
atmosphere the layer of gas that surrounds the earth
biofuel a fuel made from living things, e.g. ethanol from sugar beet, biodiesel from rapeseed oil
combustion a chemical reaction between fuel and oxygen – the burning process gives out heat
complete combustion when a hydrocarbon fuel burns with enough oxygen to form carbon dioxide and water
crude oil a natural mixture of hydrocarbons – a fossil fuel
desalination changing salt water into fresh water by separating water from the dissolved salts
fossil fuels fuels that formed millions of years ago from the remains of plants and tiny animals, e.g. coal, oil, gas
fractional distillation distillation that gives mixtures of liquids (called fractions) instead of complete separation
fractionating column a tall tower for industrial fractional distillation – the different fractions condense at different levels in the tower
global warming a rise in the average temperature of the atmosphere and the Earth's surface
hydrocarbon a compound that contains carbon and hydrogen atoms only
ignition catching fire when heated, e.g. when a flame is applied
incomplete combustion when a hydrocarbon fuel burns without enough oxygen – forms deadly carbon monoxide (and/or soot) and water
recycle reusing the same substance many times
residue the solid left after a process is complete, e.g. ash left when wood burns
sootiness the amount of soot (carbon) produced by burning fuel

sustainability being able to keep doing the same thing over and over again without harming the environment
viscosity how thick or runny a liquid is: high viscosity = thick, low viscosity = runny

C1b.8 Designer products

alcohol a type of chemical made by fermenting sugars
breathability a property of clothing that lets the water vapour from sweat escape
carbon fibre a stiff fibre made from carbon atoms only
emulsifier a chemical that stops an emulsion from separating into oil and water parts
ethanol the alcohol in alcoholic drinks
fermentation the process when yeast turns sugar into alcohol and carbon dioxide
Gore-tex™ a fabric that keeps liquid water out but can breathe
hydrophilic the part of an emulsifier that prefers water – it repels oil
hydrophobic the part of an emulsifier that repels water – it prefers oil
Kevlar™ a very strong plastic fibre – used for making bullet-proof vests
Lycra™ a very stretchy polymer (plastic) fibre
membrane a very thin layer, or sheet, of material
nanocomposite a material that has nanoparticles used in combination with other materials
nanometre one billionth of a metre $(10^{-9}$ m)
nanoparticle a particle that has a diameter between 10 and 100 nanometres
nanotechnology new technology that uses nanoparticles
smart material a material that changes its properties as conditions change
sugar a sweet-tasting carbohydrate found naturally in fruits
Teflon™ a material used to coat non-stick pans
Thinsulate™ an insulating material made of thin plastic fibres

P1a.9 Producing and measuring electricity

ammeter a meter for measuring electrical current
battery several cells connected together
capacity the amount of electrical energy a cell or battery can hold
circuit some electrical devices that have been connected together
conventional current the flow of electrical current in a circuit from positive to negative terminals
current a flow of electricity around a circuit – a number of amps (A) or milliamps (mA)
drycell a cell that produces voltage by a chemical reaction – not rechargeable
dynamo a device that generates electricity when a magnet rotates inside a coil of wire
light-dependent resistor (LDR) a resistor whose resistance changes with light intensity
magnet an object that causes a magnetic field
magnetic field the space around a magnetic object where it affects other magnetic objects
rechargeable a cell that can be used many times – a reverse current can restore its energy
resistance how difficult it is for current to flow around a circuit – a number of ohms (Ω)
resistor an electrical component that restricts the flow of electrical current – fixed-value resistors, variable resistors
series a type of circuit where all components are connected together in a single loop
solar cell a cell that produces voltage by converting light into electrical energy

superconductivity extremely low resistance that can happen at very low temperatures – electricity is conducted much better
thermistor a resistor whose resistance changes with temperature
voltage the potential difference between two points – a number of volts (V) or millivolts (mV)

P1a.10 You're in charge

earth wire the wire in a mains plug that carries current when there is a fault – green and yellow
efficiency a measure of how good something is at turning energy from one form into another useful form – a number
electricity the energy involved when charged particles flow
energy whenever something happens, energy changes from one form to another – a number of joules (J)
fuse a device in a mains plug that protects equipment and cables from excess current and the resulting fire hazard
insulation materials that can help reduce heat loss
magnetic field the space around a magnetic object where it affects other magnetic objects
motor something that turns one form of energy into movement
power how quickly something converts energy from one form to another – a number of watts (W) or kilowatts (kW)
residual current circuit breaker (RCCB) a device that shuts down an electrical circuit to protect people from electric shock
solar cell a cell that produces voltage by converting light into electrical energy
solar power generating electricity using energy that comes to us from the Sun
voltage the potential difference between two points – a number of volts (V) or millivolts (mV)
wind power generating electricity using energy from wind turbines

P1b.11 Now you see it, now you don't

absorption when a material absorbs some energy from a wave travelling in it
amplitude the maximum distance that particles in a wave move from their normal positions
analogue a signal that varies continuously, e.g. a wave
digital a signal that has only two levels, e.g. on or off
electromagnetic spectrum a group of transverse waves that all travel at the same speed in a vacuum
emission producing and sending out waves
fluorescent a substance that absorbs ultraviolet light and emits visible light
frequency how many times something vibrates in 1 second – a number of Hertz (Hz)
gamma rays ionising radiation from the highest frequency part of the electromagnetic spectrum
infrared radiation from the electromagnetic spectrum – felt as heat
longitudinal a wave where the vibration is parallel to its direction of travel
microwave radiation from the electromagnetic spectrum – used in communications
mutation when the DNA of cells is altered
optical fibre a thin strand of glass or Perspex that carries light or infra-red
radiation energy emitted from a point, spreading in all directions
reflection when waves bounce off the boundary between materials
refraction when waves enter a new material and change their speed and direction

scanning moving a detector across something to build up an image
seismic wave a wave produced by an earthquake
transverse a wave where the vibration is at right angles to its direction of travel
ultrasound sound waves with a frequency too high to hear
ultraviolet radiation from the electromagnetic spectrum – causes sunburn
vacuum a volume of space where there is no material
wavelength the distance a wave moves during the time for one full vibration
waves the way energy can be transferred by oscillations
x-rays ionising radiation from the high frequency part of the electromagnetic spectrum

P1b.12 Space and its mysteries

acceleration how quickly velocity changes
action forces come in pairs – one of the forces is called the action
asteroid a solid rock in space – about 10 to 500 km across
atmosphere the layer of gas that surrounds a planet
big bang theory suggesting that this Universe came into existence in an explosion at the beginning of time
black hole an object that has such strong gravity that light cannot escape
comet a ball of ice (and dust) in a very elliptical orbit around the Sun
dark matter material in space that is hard to detect
extraterrestrial from outside the Earth
galaxy a group of stars, e.g. the Milky Way (our Sun's galaxy)
gravitational field the space where a mass experiences a force of attraction
gravity a force of attraction that every object exerts on every other object
interplanetary the area of space where the effects of planets are negligible
mass how much of something there is – a number of kg
nebula some gas and dust in space, e.g. hydrogen
orbit movement around a point, e.g. in a circle or ellipse
oscillating theory suggests this Universe was created after the collapse of a previous Universe and that this process has happened many times
planet a natural object that orbits a star
radiation energy emitted from a point, spreading in all directions
reaction forces come in pairs – the opposing force is called the reaction
red shift the observed change in waves emitted by objects that are moving away – a decrease in frequency and an increase in wavelength
SETI (the search for extraterrestrial intelligence) an organisation that seeks life forms from other planets by searching for signals
star a huge ball of hydrogen and helium gas – it produces vast amounts of energy
steady-state theory suggests the Universe expands but does not change its appearance because new matter is being added
stellar associated with a star
Sun the star nearest to Earth
temperature a scale for measuring whether things are hot or cold
universe all of space – it includes every galaxy
weight the force caused by gravity – a number of newtons (N)
weightlessness when there is no gravitational field

Answers

B1a.1 Environment

1 Competition for resources (page 9)

1. Scavengers eat what is left after predators have eaten their prey, and other dead animals.
2. If the number of prey falls, the number of predators will fall, but this will lag behind the change in prey population.
3. Better adapted animals eat most of the food and are more likely to reproduce.
4. Animals compete for resources like food, water, territory and a mate.

2 Our influence on the environment (page 10)

1. The population of humans has increased.
2. The increase in population has put pressure on resources; forest is burned and cleared for farming, more carbon dioxide is produced by generating electricity and burning fuels, over-use of land makes it less fertile.
3. It helps to trap the Sun's radiation within the atmosphere, leading to global warming.
4. Deforestation takes place to make land for farming and to provide hardwoods for human use.

3 Chains, webs and pyramids (page 11)

1. All green plants are examples of producers.
2. The producers use energy from the Sun to produce their own food. Producers are eaten by consumers. This is how consumers get their energy.
3. A pyramid of numbers shows how many of each organism are present at each level in a food chain. A pyramid of biomass shows the dry mass of living matter at eac h level.
4. A food web is made up of many food chains linked together and gives a more complete picture of who eats what.

4 Wheat versus meat (page 12)

1. About 1% of the energy falling on a wheat field is used in photosynthesis.
2. Only about 10% of the energy eaten by the animals is transferred to humans.
3. Most energy is lost through reflection and evaporation of water from plants.
4. They are short because so much energy is lost at each stage.

5 Natural selection and evolution (page 13)

1. A fossil is an imprint of a living thing found in rocks.
2. Variations which give better adaptations to the environment give an organism more chance of surviving and breeding.
3. Organisms which cannot adapt to a big change in the environment become extinct.
4. Natural selection takes place over long periods of time and involves organisms with variations which give them better adaptations to the environment being the ones which survive and breed. Useful characteristics can be passed on to future generations and may give rise to new species.

6 Variation (page 14)

1. Humans are primates.
2. The animal kingdom is divided into vertebrates and invertebrates.
3. Some organisms display characteristics of more than one group, e.g. *Euglena*.
4. Classification is useful when new organisms are found.

B1a.2 Genes

1 Chromosomes, genes and DNA (page 16)

1. DNA
2. Brown
3. A chromosome is one, long DNA molecule, whereas a gene is a short section of DNA.
4. Alternative forms of the same gene.

2 Variation (page 18)

1. Asexual involves one parent only so offspring are genetically identical to the parent cell. During sexual reproduction the genes from two parents mix so the offspring is a variation of the parents.
2. The parent plant may have strong characteristics that the farmer wants in his new plants.
3. Sexual reproduction.
4. Our growth may be affected by our diet, plant growth may be affected by lack of minerals.

3 Inheriting disease (page 19)

1. 2
2. Mucus builds up in the digestive system slowing down the absorption of nutrients from the small intestine into the blood.
3. They are carriers of the disorder. Each parent may also contain a dominant gene that overrides the effects of the faulty allele so they appear normal.

4 Gene therapy (page 20)

1. Liposomes carrying the normal genes can be inhaled and the normal genes will be incorporated into the DNA of affected cells in the lungs.
2. A 'pattern' that your DNA forms when it is cut into random pieces. This pattern is unique to you.
3. Medical use – to find out whether a person has an inherited disease.
   Forensic use – to determine whether someone is guilty of a crime.

5 The human genome project (HGP) (page 21)

1. A project working on human DNA. Scientists have 'mapped' or worked out where each of our genes are on a chromosome and on what chromosomes they are found.
2. We could decide what treatment a person needs or we could eventually replace the faulty gene with a healthy one.
3. Advantages
1) earlier diagnosis of genetic disease 2) it will help gene therapy in that the right treatment can be organised.

Disadvantages

1) the 'wrong' people might get hold of the information and use it against us e.g. insurance companies 2) who decides that a test should be carried out e.g. a doctor or an employer?

6 Playing with genes (page 23)

1. Antibodies, hormones
2. Advantages
1) producing individuals with strong characteristics 2) allows an infertile couple to have children.

Disadvantages

1) does not allow natural evolution so organisms can not adapt to environmental changes 2) may encourage individuals to clone particular characteristics to create, for example, a race of superhumans.

3. Designer milk is normal milk that has been changed or redesigned through genetic engineering to contain a new substance.

Designer babies may be created so a parent can choose the characteristics of their child so that the child has, for example, a certain eye or hair colour.

B1a.3 Electrical and chemical signals

1 The nervous system (page 25)

1. brain, spinal cord
2. It coordinates or sorts out incoming information and decides what electrical signals need to go out and where they need to go.
3. sensory neurones.
4. The electrical signals in the brain are not 'ordered' and sudden bursts of electrical activity that occur randomly can cause unconsciousness and seizures.

2 Sense organs (page 26)

1. The part of a sense organs that detects a change inside or outside of the body.
2.
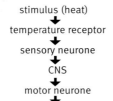
stimulus (heat)
↓
temperature receptor
↓
sensory neurone
↓
CNS
↓
motor neurone
↓
effector (muscle to move hand away)
3. Brightness of light (light intensity)

3 The eye (page 28)

1. When the lens changes shape to ensure that light is focussed at a particular point on the retina at the back of the eye.
2. Retina.
3. When the size of the pupil is adjusted by contraction of the iris muscles.
4. Motor

4 Reflex and voluntary actions (page 29)

1. They help to protect our body from harm.
2. Accommodation and iris reflex.
3. Brain and head
4. They do not involve the brain.

5 Chemical messengers (page 31)

1. A chemical messenger
2. Insulin
3. Plasma
4. Less side effects and can be produced in large amounts.

6 Fertility and infertility (page 32)

1. Oestrogen and progesterone
2. A drug that contains sex hormones to prevent fertilisation of the egg and therefore pregnancy.
3. In-vitro fertilisation is when an egg is fertilised in special conditions outside of the body.
4. Advantage – Allows childless couples to have children.
   Disadvantage – destroys embryos that are not needed.

B1a.4 Use, misuses and abuse

1 Tuberculosis (page 33)

1. An antibiotic is a drug used to kill bacteria inside the body.
2. If people do not complete a course of antibiotics then they risk not killing all the harmful bacteria. If the only bacteria that survive are the ones that are harder to kill, then they will pass that characteristic onto their offspring. If this process continues

for several generations the bacteria may develop resistance to the drug.
3. Many AIDS sufferers die from TB because the HIV virus destroys the immune system, so the body has no defence against TB bacteria.
4. New drugs cost a lot of many to develop because so much testing must be done and this can take ten to fifteen years.

2 Microorganisms and disease (page 34)

1. Microorganisms are removed from the breathing system by cilia lining the breathing system, after the microorganisms have become trapped in mucus.
2. The second line of defence against microorganisms is called the non-specific immune system because white blood cells react the same way to all pathogens, ingesting and destroying them
3. Vaccination gives you immunity because you are injected with a small harmless amount of infection. Your immune system creates antibodies which remain in the body in case the real pathogen invades
4. DOTS means that a nurse watches the patient swallow every tablet to make sure they complete their course of treatment. This reduces the risk of creating drug resistant bacteria.

3 Types of drugs (page 35)

1. Alcohol and barbiturates are examples of sedative drugs.
2. Stimulants can improve reaction time because they increase the speed of the nervous system.
3. Doctors are very careful about prescribing barbiturates for epilepsy because the drugs can be addictive.
4. Stimulants can increase the rate at which impulses cross synapses, sedatives slow them down or even stop them.

4 Pain relief (page 36)

1. Paracetamol, morphine and heroin are all painkillers.
2. Painkillers affect the nervous system by blocking some of the impulses that carry information about pain to the brain.
3. Morphine and heroin are very dangerous because they cause addiction and tolerance.
4. Too much paracetamol can damage the liver and stop it from working.

5 Drug misuse and abuse (page 37)

1. Solvent abuse slows down breathing and heart rate. It also causes nausea, headaches and blackouts, and can cause heart failure.
2. Solvents can damage the myelin sheath of neurones, making it harder for them to carry impulses
3. Alcoholism is when a person has become dependent on alcohol. They need more to get the same effect.
4. It is dangerous to drink and drive because alcohol slows down your reactions, so you are more likely to have an accident

6 Tobacco (page 38)

1. Nicotine, carbon monoxide and tar are found in cigarettes and can affect the body
2. Smoking affects your heart because nicotine speeds up your heart, but also narrows your blood vessels so increasing your blood pressure. This may cause heart disease.
3. Long-term smokers have to have extra oxygen because the air sacs in the lungs have been weakened so that there is less surface area for gaseous exchange.
4. Bronchitis causes the air passages to become inflamed and the cilia lining these stop beating. Mucus, dirt and bacteria stay in the lungs.

C1a.5 Patterns in properties

1 The periodic table (page 40)

1. (a) Cl   (b) Co   (c) C   (d) Ca
2. (a) neon   (b) sodium   (c) nitrogen   (d) nickel
3. Co is the metal cobalt and CO is the compound, carbon monoxide, formed from carbon and oxygen.

2 The atom (page 41)

1. (a) 20   (b) 8   (c) 36
2. (a) gold   (b) lead   (c) silver

3 Analysis and identification (page 42)

1. (a) lithium   (b) barium   (c) sodium
2. (a) nickel   (b) copper   (c) iron(III)
3. dark green

4 Alkali metals (page 43)

1. Sodium + water → sodium hydroxide + hydrogen
2. BaNaNa (Banana)

5 Halogens (page 44)

1. Astatine, iodine, bromine, chlorine, and fluorine
2. Solid at room temp, very dark colour, metallic

6 Noble gases (page 44)

1. Because they have a complete outer shell and therefore do not need to lose, gain or share electrons.
2. (a) Helium, party balloons, airships and scuba tanks
   (b) Argon, electric light bulbs and fluorescent tubes

C1a.6 Making changes

1 Oxygen (page 45)

1. Oxygen is needed for burning, with more oxygen there is more burning.
2. Magnesium + oxygen → magnesium oxide
   $2Mg + O_2 \rightarrow 2MgO_3$.
3. Because oxygen is removed during the reaction.

2 Other important gases (page 46)

1. Carbon dioxide and water
2. Carbon dioxide because both carbon and oxygen are left over in this reaction
3. Collected by upward delivery because the gas will rise up into the test tube. Ignites with a squeaky pop.

3 Metals and reactivity (page 47)

1. Carbon dioxide
2. Reduction
3. Calcium oxide and copper

4 Metals and their salts (page 48)

1. (a) magnesium + sulphuric acid
   magnesium sulphate + hydrogen
   $Mg + H_2SO_4 \rightarrow MgSO_4 + H_2$
   (b) zinc + sulphuric acid → zinc sulphate + hydrogen
   $Zn + H_2SO_4 \rightarrow ZnSO_4 + H_2$
2. $2Na + Cl_2 \rightarrow 2NaCl$
3. Metal hydroxides contain oxygen and hydrogen in addition to the metal. When they react with an acid, a further hydrogen atom is gained, therefore water is formed.

5 Common compounds in foods (page 49)

1. The biggest problem is tooth decay.
2. sodium hydrogen carbonate → sodium carbonate + carbon dioxide + water
3. Thermal decomposition
4. Causes bubbles to be formed which lead to the cake rising

6 Chemicals for cleaning (page 50)

1. Ammonia, calcium chloride and water
2. Corrosive
3. They increase the amount of available oxygen for burning, therefore it burns brighter.

C1b.7 There's one Earth

1 Early Earth to the present day (page 51)

1. Carbon dioxide
2. Carbon dioxide
3. Fractional distillation
4. Photosynthesis by prehistoric plants

2 Global warming (page 52)

1. If we are uncertain of the effects of something we should act to prevent it. If the effects of an excess of carbon dioxide in the atmosphere are unknown, we should act to stop the carbon dioxide being produced.
2. Sea level rise; Size of polar ice sheets; Salinity of sea water
3. Less carbon dioxide is removed by photosynthesis

3 Crude oil and fuels (page 54)

1. The plants remove carbon dioxide from the atmosphere whilst growing, which is released back to the atmosphere when they are burned as fuels.
2. Viscosity decreases as you move up the fractionation column, as does the length of the carbon chains.
3. 2 (C_2H_4 and C_5H_{10})

4 Combustion reactions (page 55)

1. $2C_2H_6 + 7O_2 \rightarrow 4CO_2 + 6H_2O$
2. Carbon monoxide bonds with red blood cells, which stops oxygen bonding to them and therefore the body is unable to get oxygen.
3. Oxygen, it is essential for burning

5 Recycling (page 56)

1. Recycling uses 20% less energy than making new glass and we do not use up the Earth's natural resources.
2. There are no fuel costs. It is better for the environment; no pollution.
3. Sustainable development means there will be a good environment for future generations.

6 The use of sea water (page 57)

1. Chlorine forms negative ions which are attracted to the positively charged anode.
2. Sodium stays in solution as aqueous sodium hydroxide.
3. Soaps, detergents and oven cleaners.

C1b.8 Designer products

1 Smart materials (page 58)

1. Kevlar™ is a lot lighter, and therefore easier to wear as part of a uniform.
2. Teflon™ and the adhesive on Post-it™ notes.

2 Nanoparticles (page 59)

1. Sunscreens to reflect UV, silver nanoparticles to kill bacteria on toothbrushes, dummies and medical equipment, make fabrics stain resistant, stop organic particles sticking to glass
2. Material that uses nanoparticles in combination with other materials

3 Fermentation (page 60)

1. Yeasts are living things that rely on enzymes. Enzymes work best when it's warm.
2. Either the food (sugar) supply will run out or the build up of toxic alcohol kills the yeast.

4 Intelligent packaging (page 61)

1. The nitrogen prevents the crisps going soft and mouldy, but also acts as a buffer to stop them being squashed. Removing the air would prevent them going soft and mouldy but would make them easy to damage in transit.
2. Food is kept microbe free but nutrients etc may be breaking down. Just because the food looks fresh doesn't mean it is fresh.

5 Emulsions (page 62)

1. A chemical in egg yolk that acts as an emulsifier, in mayonnaise.
2. The hydrophobic ends of the detergent attach to oil on cloths, the hydrophilic ends stick to the water. Droplets of oil are then pulled into the water to form an emulsion and leave your clothes clean

6 Designer products (page 63)

1. removes dirt and oil from hair, rinses easily, and doesn't damage hair
2. smells nice, eye catching colour, lathers well

P1a.9 Producing and measuring electricity

1 Producing electric current (page 64)

1. A device that generates electricity when a magnet rotates inside a coil of wire.
2. d.c. flows in one direction only, a.c. flows first one way and then the other.
3. (a) There is more current so the lamps are brighter.
   (b) There is no current so the lamps go out.
4. Slow the dynamo down .
   Use a weaker magnet .
   Use fewer turns of wire on the coil.

2 Current, voltage and resistance (page 65)

1. the ampere (A)
2. the ohm (Ω)
3. (a) the current increases
   (b) the electricity has more energy
4. $6 = 2 \times R$
   $R = 6 \div 2 = 3\ \Omega$

3 Resistance, lamps and computers (page 66)

1. An electrical component that restricts the flow of electrical current.
2. Your graph should have a straight line that slopes upwards.
3. It increases.
4. For instance: at 3V, the resistance is $3 \div 0.78 = 3.8\ \Omega$
   and at 12V, the resistance is $12 \div 1.62 = 7.4\ \Omega$
   The resistance increases as the voltage increases.

4 Changing resistance and controlling the current (page 67)

1.

2. A resistor whose resistance changes with temperature.
3. For instance, digital thermometer, computer cooling fan control, central heating thermostat.
4. A LDR is connected in series with the clock display.

5 Cells and recharging (page 68)

1. Several cells connected together (usually in series).
2. A cell that can be used many times – a reverse current can restore its energy.
3. Batteries contain poisonous chemicals which can be reused.
4. Time = capacity ÷ current = 900 ÷ 45 = 20 hours

6 Electricity and technology (page 69)

1. For instance, by writing a letter.
2. Heat, light, sound, etc.
3. It would need:
   – superconducting electromagnets to provide the lifting force
   – magnetic traction to pull it along and to slow it down
   – special track to keep it going in the right direction.

P1a.10 You're in charge

1 Renewable energy v. fossil fuels (page 70)

1. Coal, gas, oil
2. Renewable resources (like energy from the Sun) do not run out, non-renewable resources (like energy from oil) are finite and will run out eventually.
3. When coal is burned, gases that pollute the air are produced.
4. The wind does not always blow.
5. Look at the results of a survey that collected data about a particular hazard (e.g. people getting ill) for different distances from the cables.

2 Motors (page 72)

1. For example, DVD player, vacuum cleaner, washing machine.
2. Reversing the direction of the magnetic field or reversing the direction of the current.
3. The strength of the magnetic field, the size of the current and the number of turns on the coil.
4. Because the current from the battery is d.c. (Also d.c. is easy to reverse when the car needs to go backwards.)

3 Wires, fuses and safety (page 73)

1. A device in a mains plug that protects equipment and cables from excess current and the resulting fire hazard.
2. The fuse will blow if the current exceeds 5A.
3. (a) The fuse and earth wire should be enough to prevent shock.
   (b) If you use a RCCB, it should stop the current if you get a shock.
4. The drill was earthed. The drill had a fuse. The lights had a fuse.

4 Power (page 74)

1. How quickly something converts energy from one form to another.
2. The units are watts (W) or kilowatts (kW).
3. Power = 5 × 230 = 1150 W (or 1.15 kW).
4. Current = 2300 ÷ 230 = 10A
   The current in the kettle is 10A, so it needs a fuse labelled 10A, or a bit more.
   A 13A fuse would be fine.

5 Paying for electricity (page 74)

1. The kilowatt-hour (kWh).
2. The smaller amount of water takes less time to boil, so it needs less energy.
3. cost = power × time × cost of 1 kWh = 3 × 6 × 12 = 216p
   (or £2.16)
4. pay-back time = cost ÷ yearly saving = 6000 ÷ 30 = 20 years

6 Solar cells and efficiency (page 75)

1. A measure of how good something is at turning energy from one form into another useful form. It has no unit because it is a number or a percentage.
2. Compared to ordinary bulbs, low energy light bulbs turn more energy into light and less into wasted heat.
3. Efficiency = useful output ÷ total input (× 100 %) = 40 ÷ 100 = 0.4 (or 40 %)
4. Rarely used at home – expensive, do not provide AC, cloudy days reduce power output.
   Often used for satellites – lightweight, provides DC to charge batteries, the Sun shines brightly in space.

P1b.11 Now you see it, now you don't

1 Wave basics (page 76)

1. Sound, ultrasound or some seismic waves.
2. For the same frequency, its speed increases.
3. They both have exactly the same speed in a vacuum although the speed of microwaves will be very slightly slower in air (remember refraction of light).
4. For example, ultraviolet has a higher frequency but a smaller wavelength than infrared but both travel at the same speed in a vacuum.

2 Reflections (page 77)

1.

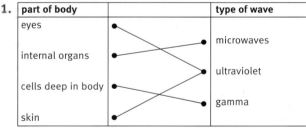

2. total internal reflection (TIR)
3. Laser light shines on the barcode. The black parts absorb the light and the white parts reflect it.
4. 333 m/s (the sound travels 2 x 100 m so use 200/0.6)

3 Scanning by absorption and emission (page 78)

1. The most obvious one is bone, but other materials like metals do as well.
2. Emission is when radiation is given out. Absorption is when the radiation is taken in.
3. Think about an electric fire. It takes some time before it glows. It first produces infrared radiation (heat) and then moves through the red, orange and yellow parts of the visible spectrum.
4. The energy for X-ray photographs is absorbed by the bones, etc. Ultrasound is reflected. Absorption is the dangerous part.

4 Digital signals and mobile phones (page 79)

1. (a) analogue
   (b) digital
2. ultraviolet and some X-rays

5 How dangerous are these waves? (page 80)

1.
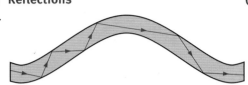

part of body		type of wave
eyes		microwaves
internal organs		ultraviolet
cells deep in body		
skin		gamma

2. X-rays have a much higher frequency and so have greater energy which has to be absorbed.

6 Earthquake waves (page 81)

1. (a) Primary- and Secondary-waves
   (b) Primary are quicker and longitudinal, while secondary are transverse.
2. A seismic wave is inside the solid crust of the Earth. Tsunamis are in/on the water.
3. Only P waves would be transmitted to the south pole since transverse (S) waves cannot pass through liquids.

P1b.12 Space and its mysteries

1 A weighty problem (page 82)

1. mass
2. just over 2 million newtons

3.

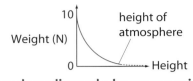

2 Where do we live and who are our neighbours? (page 84)

1. Mars
2. (a) Jupiter has nearly [[3/4]] of the mass of all the planets.
   (b) They have too little mass to be seen on the chart.
3. The orbits of planets are mainly circular, while those of comets are long and thin. The planets orbit in one plane while the comets can orbit in any plane.
4. That depends on you. But as a scientist, you might want to know things like:
   – where is it?
   – how big is it?
   – how far from its star is it?
   – how did we find out about it?
   – might it support life? etc.

3 I'm an Earthman – get me out of here (page 85)

1. newtons or N
2. Fire your jet in the opposite direction to the way you want to go.
3. They have different masses. The lighter one will accelerate more.
4. 6.9 million N (6.9 MN) (F = ma, so F =2.3 x 3)

4 What's it like out there? (page 86)

1. much higher because the force of gravity is less and there is no air resistance, but you may not be sure because you will also need a space suit and oxygen tanks, etc.
2. (a) 20 kg   (b) 20 kg (c) 460 N
3. (a) P
   (b) 6 years (3 orbits for Q and two orbits for P)
   (c) 3 years
4. e.g.temperature, atmosphere, radiation and gravity

5 Is anybody else there? (page 87)

1. Search for ExtraTerrestrial Intelligence.
2. Can make decisions/react to unexpected situations.
3. Do not need:
   – food
   – water
   – atmosphere
   – radiation control
   – accurate temperature control
   – beds/material for entertainment
   – room for moving
   – artificial gravity
4. 50 years (i.e. 50 there and 50 back! = 100)
5. This of course depends on you! You might be frightened and go to hide. Alternatively, you might want to zap them or greet them as friends and try to learn from them.

6 Where do we go from here? (page 88)

1. The Universe is expanding up to a maximum but at a slower pace then it starts to contract at a gradually increasing pace until it becomes zero. Then it starts to increase rapidly again. This is the Oscillating Universe theory.

2.
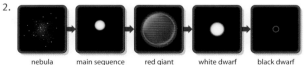

nebula   main sequence   red giant   white dwarf   black dwarf

3. The stars all around us seem to be moving away from us, so the Universe must be getting bigger.